Excel

SUCCESS ONE® HSC

BIOLOGY

Past HSC questions
2001–2014 by Module
2015–2018 by Year

Past 2019–2022 HSC papers

PASCAL PRESS

ISBN 978 1 74125 744 1

Pascal Press
PO Box 250
Glebe NSW 2037
(02) 8585 4050
www.pascalpress.com.au

Publisher: Vivienne Joannou
Commissioning and project editor: Mark Dixon
Edited by Karen Enkelaar and Rosemary Peers
Answers checked by James Mansfield, Diane Alford, Sarah Barrand, Jennifer Hill and Sonia Sharma
Cover design by Michael Sherman
Typeset by Grizzly Graphics (Leanne Richters)
Cover photo by The *Sydney Morning Herald*
Printed by Vivar Printing/Green Giant Press

Foreword

Congratulations on selecting Biology as part of your HSC program of study. This course will prepare you well for further tertiary study or a career in academia. It will also give you a lifelong understanding of the key principles and thinking skills of Biology.

Excel *Success One HSC Biology* is a valuable learning tool that has been developed to help you with your HSC preparation. This edition has been kept up to date with the inclusion of the 2022 HSC Examination paper with sample answers. Your teacher will have many suggestions for how to use this resource to help support your preparation for the HSC Examination.

This edition also contains selected questions from past HSC papers which are relevant to the new Year 12 Biology syllabus. We have also included extra questions written specifically for the new content with sample answers provided.

On behalf of the Science Teachers' Association of New South Wales we hope that your year will be enjoyable and productive, and the results a just reward for your efforts.

STANSW thanks those members who continue to be involved in the production of this book.

Margaret Shepherd
STANSW President
January 2023

The Mark Maximizer Guide was written by Diane Alford.

HSC EXAMINATION PAPER QUESTIONS

SAMPLE ANSWERS

The sample answers to HSC questions contained in this publication are examples of answers that the authors believe would score full marks. However, they are not endorsed by NESA.

Contributors to the sample answers include Diane Alford, Jennifer Hill, Monika Khun, Mary Cahill, David Randall, Dianne Shaw and Col Harrison.

COMMENTS

The Science Teachers' Association of NSW (STANSW) and Pascal Press welcome constructive comments on the questions and answers for future editions of this book.

FORMAT

This guide's format has been constructed to support students' revision for the HSC Examinations based on the Biology Stage 6 syllabus 2017 (updated January 2018):

- The **first section** contains objective-response and short-answer questions from past HSC Examination papers (2001–2014) as well as sample questions. This section is arranged according to the syllabus Module topics.
- The **second section** contains four sample HSC Examination papers with actual questions from the 2015–2018 HSC papers. All of the questions in these papers are on topics examinable in the new syllabus. Approximately half of each paper's questions are actual HSC questions and the others are questions on the new topics in the updated syllabus.
- The **third section** consists of the 2019–2022 HSC Examination papers.

Contents

Mark Maximizer Guide

This Mark Maximizer Guide will arm you with strategies and tips so you can maximize your marks by making every minute of your HSC Examination count. And it has some great tips for your exam preparation too. Read it. Use it.

The structure of the HSC Examination paper

All Science exams consist of two sections. The following table summarises the structure, which changed in 2019. All questions are compulsory.

Section	*Marks*	*Question type*	*Answers*	*Recommended time*
I	20 (1 mark per question)	Objective response	Fill in the responses on the answer sheet.	35 minutes
II	80 (varying mark values with at least two items worth 7–9 marks)	Short answer: Questions may have parts. There will be at least 20–25 items. A variety of stimulus material such as text, diagrams, pictures, graphs, photographs and illustrations that are essential for answering the question may be included.	Answer questions in the spaces provided, showing all working and calculations. Space provided is a guideline for expected answer length but extra space is provided at the back of the booklet. If you use extra space make sure you clearly indicate the question number or part. Some answers may require a diagram or graph while in other answers you may choose to support it with diagrams or other graphical material.	2 hours 25 minutes

Time allocation

The examination begins with 5 minutes reading time. You will not be able to write in this time. Rather than starting to read the objective-response questions at the beginning, you could:

- spend this time reading some of the short-answer questions in Part B and Section II,

  or
- select some of the easy questions to attempt earlier in the exam after completing the objective-response questions, to build your confidence for the more difficult questions later.

Time management

A planned approach to time management in the exam is important. There are several strategies you could use:

- Prepare an exam timetable
  The table above shows the suggested times for each section as recommended by NESA. If you choose to proceed through the exam in an orderly sequence of sections or parts, you can work

out an exam timetable before the exam. You will then know when you should be changing sections and check that you are on schedule. Try to be a bit ahead of schedule so that you have time to check answers, make corrections or return to more difficult questions.

- Calculate time allocated per mark
  If you do not choose to work through the exam in a fixed sequence, you may wish to calculate time allocations for questions based on the 100 marks and 180 minutes in which to record answers. Each mark should take 1 minute 48 seconds to earn. The following table gives calculations of approximately how long you should spend on questions of a particular mark value.

Question value (marks)	*Suggested time*
1	1 min 45 s
2	3 min 30 s
3	5 min 20 s
4	7 min 00 s
5	9 min 00 s
6	10 min 45 s
7	12 min 45 s
8	14 min 30 s
9	16 min 00 s
10	18 min 00 s

- Don't panic!
  Worrying about time too much will only add to your stress and waste time. There is no need to time yourself for each question, and particularly not for the low-value questions. However, if you panic or lose track of time or waste too much time on a question of relatively small mark value, you may be severely disadvantaged.

Analysing the questions

Questions in the HSC are designed to enable students to show their achievement of a range of outcomes over a range of levels. They are not trying to trick you, and the question will state clearly all that is required of you. If marks are to be awarded for giving an answer in the correct units, then the question will clearly state that you need to give the answer in the correct units.

- Highlight key terms
  Highlight or underline the key terms or things you have to do (the verbs) and the key ideas or concepts that make up the question—a useful strategy.

- Learn the lingo
  The glossary of terms that are often used in questions will help you analyse what is required of the question (see the Glossary of key words on page ix).

- Remember that questions are planned to vary in their level of difficulty
  Don't panic or waste excessive time on a question that you find particularly difficult. The following table may give you some ideas on how to find the easier questions that you can attempt first. Then, when you have finished everything that you can do, you can move on to spend time with the very difficult questions.

Question type	*Key words that indicate level of difficulty*
Difficult or complex questions often demanding high levels of skills to use the facts	assess, construct, critically analyse, critically evaluate, evaluate, justify, propose, recommend
Questions of medium levels of difficulty requiring some thinking and using the facts	account for, analyse, apply, assess, calculate, classify, compare, contrast, deduce, demonstrate, discuss, examine, explain, estimate, extract, extrapolate, how…, interpret, investigate, predict or recommend, sketch, why...
Simple or direct questions of low levels of difficulty based on remembering and communicating the facts	account, clarify, define, describe, identify, outline, recall, recount or summarise, state, what..., which...

  Also, there may be variation in difficulty within the topics you have studied. A relatively easy remembering-type question may become a little more difficult if it is about a particularly difficult idea.

- Look at question structure and mark value
  The structure of the question may also contribute to differences in levels of difficulty. The mark value for the question may give you some clues. For example if the question is only two lines long but is worth six marks, that is a clue that it may require extra care to ensure all aspects of the question have been covered.

  Many students find the questions that are divided into subsections easier to attempt. They feel that each small portion of the question can be 'bitten off' at a time. Often the answer to one subsection helps to provide some ideas to help with the other subsections.

  The more difficult integrated questions may seem a bit more difficult to swallow, and when you do, they may give you indigestion. Practising these questions using examples from this book will really help you develop your skills and help avoid the need for antacid tablets!

- Answer all parts
  A common error is that students often only answer one part of a two-part question. This is very easy to do if the parts are not clearly numbered but are written as sentences following each other. It is worth making a note on the paper of any multiple-part questions and make sure you return to the question to finish all sections.

Glossary of key words

For each subject you should prepare summaries and a glossary of the key definitions. You cannot answer questions if you are unfamiliar with the content of the subject.

You also need to be familiar with the language of exam questions. This involves understanding the verbs that are the key words that tell you what you have to do to earn the marks. Below is a table of the key verbs that may be contained in questions and the definitions of those words.

Key word	*Definition*
Account	Account for: state reasons for; report on Give an account of: narrate a series of events or transactions
Analyse	Identify components and the relationship among them; draw out and relate implications
Apply	Use, utilise, employ in a particular situation
Appreciate	Make a judgement about the value of
Assess	Make a judgement of value, quality, outcomes, results or size
Calculate	Ascertain/determine from given facts, figures or information
Clarify	Make clear or plain
Classify	Arrange or include in classes/categories
Compare	Show how things are similar or different
Construct	Make; build; put together items or arguments
Contrast	Show how things are different or opposite
Critically (analyse/ evaluate)	Add a degree or level of accuracy, depth, knowledge and understanding, logic, questioning, reflection and quality to
Deduce	Draw conclusions
Define	State meaning and identify essential qualities
Demonstrate	Show by example
Describe	Provide characteristics and features
Discuss	Identify issues and provide points for and/or against
Distinguish	Recognise or note/indicate as being distinct or different from; to note differences between
Evaluate	Make a judgement based on criteria; determine the value of
Examine	Inquire into
Explain	Relate cause and effect; make the relationships between things evident; provide why and/or how
Extract	Choose relevant and/or appropriate details
Extrapolate	Infer from what is known
Identify	Recognise and name
Interpret	Draw meaning from

Key word	*Definition*
Investigate	Plan, inquire into and draw conclusions about
Justify	Support an argument or conclusion
Outline	Sketch in general terms; indicate the main features of
Predict	Suggest what may happen based on available information
Propose	Put forward (for example, a point of view, idea, argument, suggestion) for consideration or action
Recall	Present remembered ideas, facts or experiences
Recommend	Provide reasons in favour
Recount	Retell a series of events
Summarise	Express, concisely, the relevant details
Synthesise	Put together various elements to make a whole

Answering the question

Objective-response questions

Many people find objective-response (multiple-choice) questions relatively easy. Recording the answer is certainly easy, but they can be deceptive in their level of difficulty. Some make up for the time saved recording the answer by requiring time to think about the answer. Some of the distracters (i.e. the wrong answer options) may be based on common mistakes or misconceptions and may lead you away from the correct answer. In one 2000 HSC Science exam only 15% of students got one particular objective-response question correct, showing how effective the distracters can be.

Careful reading and interpretation of questions are critical. Sometimes you may be left with two similar answers and you have to dig into your deeper understanding to work out the implications of the subtle difference. The more practice you get on objective-response questions, the more skilful you will become. In this book we have provided explanations for making the correct choices; you do not have to do this in an exam.

Short-answer questions

- Highlight the verb and analyse the question
  After you have analysed the question, link it to the specific module and the concepts of that module. If the question is in parts, look at the parts and see if they give you clues about the other sections of the question. If the question gives you the freedom to select the format of your answer, choose a format that you feel most comfortable with. For example, you may prefer to express some ideas in a table, flowchart or diagram—or in writing. This may be especially important in those questions with a high mark value that require you to integrate ideas.

  Use the mark value and space provided to give you a rough guide to the number of points in the answer and the length of the answer. If it is worth six marks, you may need to include at least six key points or one point with a set of arguments, supportive statements, reasons or counterarguments.

- Read through your answers
  It is important to read through your answers to short-answer questions to check that you have clearly communicated your understanding. Check that any assumptions you have made have been stated and that your thinking is clear. If you are running out of time it is important to attempt questions in a very concise manner; even if you remain uncertain as to what the question is about, think back to the option and try to make some links to key ideas.
- Attempt all questions
  It is essential you attempt all questions. You do not lose marks for guessing or recording wrong answers. Remember, however, if you ramble on with a lengthy, confused or contradictory answer you may make a satisfactory answer meaningless and therefore worth no marks. Clear and concise answers are always preferable. If you are running out of time, writing key points, tables or diagrams may help communicate an answer or earn partial marks.

  Avoid persevering with one question to make it perfect if it means you leave out several others. The mark value of questions can serve as a guide for prioritising your attempts at questions when running out of time. You will do yourself a great disservice if you do not attempt questions worth six or more marks.
- Show your working

Common mistakes

Examiners provide reports after each HSC Exam that indicate areas of strengths and weaknesses in student responses. From these common mistakes, it is clear that students generally can benefit by:

- more precise and scientific use of terminology, especially that specified in the syllabus
- more precision with graphs and labelling of diagrams
- avoiding answers that are too long or detailed or that use ambiguous terms
- correct interpretation of questions, e.g. listing instead of describing, or giving a description not an explanation
- answering in specific terms not generalities
- ensuring all parts of multi-part questions are attempted
- being able to clearly describe procedures, apparatus, etc. for mandatory practical activities as a firsthand experience.

Golden rules

Preparing for the exam

- Make certain you have packed all the equipment you need, such as the correct calculator, ruler and pencil for diagrams.
- Obviously the exam will reward those students who have a thorough knowledge of the whole course. Do all assignments and assessment tasks to your best ability and use them in your revision.
- Get as much practice at completing and getting feedback on exam-type questions as you possibly can. This book is perfect for this!
- Make certain you complete, know and revise the mandatory practicals from the syllabus so you can describe the procedure or evaluate its methods.

In the exam

- Don't give up! If you find it difficult, so will most other students. Attempt all questions with concise, clear answers.
- Have a plan that ensures you manage time well. If the option is your particular strength, you may start with it, but do not spend more than 45 minutes on it. Alternatively, start with the objective-response because the correct answer must be there, and it can help you get over your nerves.
- Know yourself and work to your strengths. If you want to experiment with different strategies for tackling the exam, do that in the trial exam and reflect on how the plan worked.
- At the end re-read the questions and the answers you have written. Check that you have really communicated what you intended to write.
- Don't panic! It can lead you to make some silly decisions. There will be some difficult questions in the exam but there will also be some easy ones.

Finally, GOOD LUCK!

CHAPTER 1

Module 5
Heredity

Part A Objective-response questions

1 The graph shows information about four species of bacteria and their reproductive rates at different temperatures.

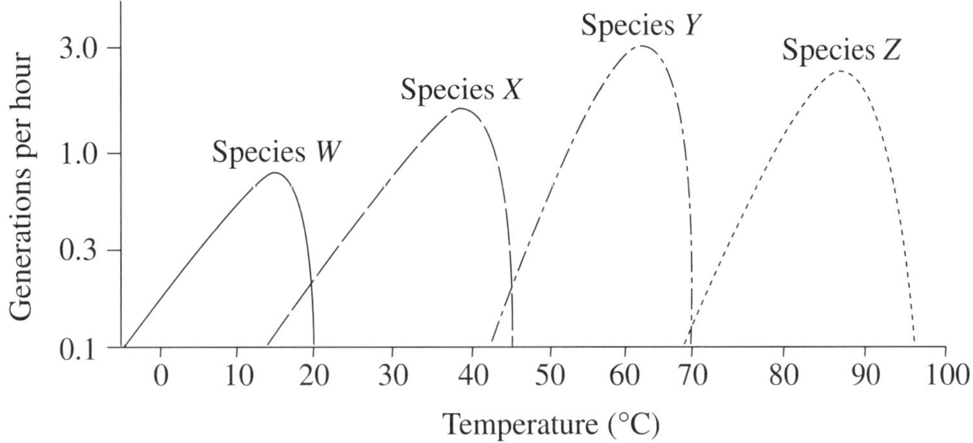

What conclusion can be drawn from this graph?

(A) All bacterial species can adapt to a broad range of temperatures.

(B) Individual species can reproduce in a broad range of temperatures.

(C) All bacterial species are limited to a range between 0 °C and 100 °C.

(D) Individual species reproduce in a relatively narrow range of temperatures.

(q3, 2002 HSC)

2 The diagram represents one pair of homologous chromosomes during meiosis.

Crossing-over occurs and random segregation takes place.

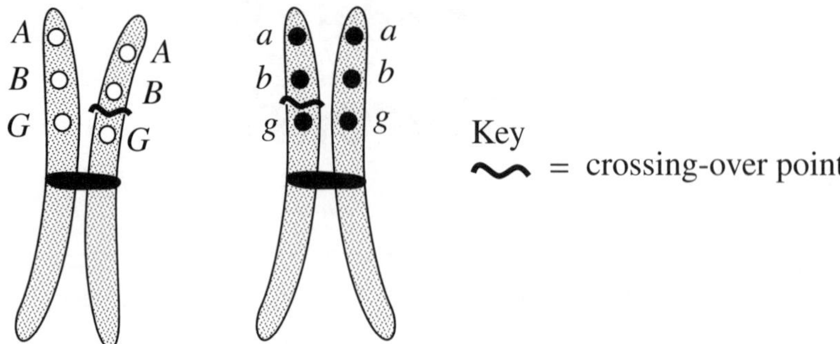

What genotypes are produced?

(A) *ABG*, *abG*, *ABg*, *abg*

(B) *ABG*, *aBG*, *Abg*, *abg*

(C) *ABG*, *ABG*, *abg*, *abg*

(D) *ABG*, *aBg*, *Abg*, *abg*

(q10, 2001 HSC)

3 In a variety of garden peas, the allele for tall plants (*T*) is dominant over the allele for short plants (*t*). A cross between a tall plant and a short plant resulted in 50% of the offspring being short.

What were the genotypes of the parents?

(A) *Tt* and *tt*

(B) *Tt* and *Tt*

(C) *TT* and *Tt*

(D) *TT* and *tt*

(q1, 2002 HSC)

4 In rabbits, a black (B) coat colour is dominant over a white (b) coat colour.

Which Punnett Square correctly represents a cross between a rabbit heterozygous for coat colour and a white rabbit?

(A)

	B	b
B	BB	Bb
b	Bb	bb

(B)

	B	B
b	Bb	Bb
b	Bb	Bb

(C)

	B	b
b	Bb	bb
b	Bb	bb

(D)

	B	b
b	Bb	Bb
b	Bb	bb

(q13, 2004 HSC)

5 Which of the following statements best describes the process of hybridisation frequently used in agriculture?

(A) The transfer of a gene from one species to another

(B) The crossing of two genetically different strains of a species

(C) The production of genetically identical offspring by cloning

(D) The artificial selection and breeding of suitable offspring within a species

(q13, 2002 HSC)

6 The steps involved in DNA replication and protein production are summarised below.

In which step would a mutation lead to the formation of a new allele?

Step *A*:	DNA is copied and each new cell gets a full copy.	DNA replication
Step *B*:	Information is copied from DNA and taken to cytoplasm.	Protein synthesis
Step *C*:	Ribosome reads information and assembles protein.	
Step *D*:	Protein formation is completed.	

(A) Step *A*

(B) Step *B*

(C) Step *C*

(D) Step *D*

(q15, 2002 HSC)

7 How are new alleles formed?

(A) By crossing over during meiosis

(B) By cloning a new variety in a population

(C) By mutation in the DNA of a gene

(D) By the production of a new phenotype from the same DNA

(q1, 2009 HSC)

8 Thirty per cent (30%) of the nucleotide bases in human DNA are adenine (A).

What is the percentage of guanine (G) bases in human DNA?

(A) 20%

(B) 30%

(C) 40%

(D) 70%

(q7, 2009 HSC)

9 The gamete plays an important role in sexual reproduction because it carries:

(A) genetic information from both parents.

(B) half the genetic information of the parent.

(C) all of the genetic information of the parent.

(D) double the genetic information of the parent.

(q5, 2010 HSC)

10 Which of the following shows DNA replication in the correct order?

(A) Two DNA double helices → strands separate → matching bases pair up → DNA double helix

(B) DNA double helix → strands separate → matching bases pair up → two DNA double helices

(C) Strands separate → two DNA double helices → matching bases pair up → DNA double helix

(D) DNA double helix → strands separate → two DNA double helices → matching bases pair up

(q13, 2010 HSC)

11 Consider the following pedigree chart.

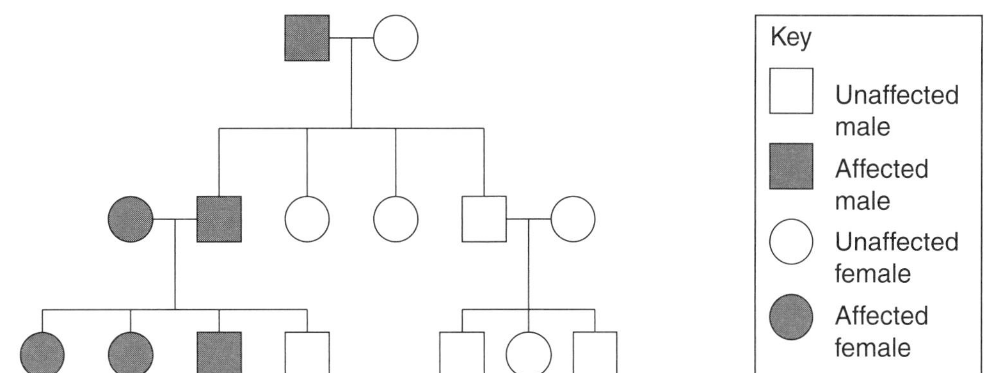

What type of inheritance is shown in the pedigree?

(A) Sex-linked recessive

(B) Sex-linked dominant

(C) Non–sex-linked recessive

(D) Non–sex-linked dominant

(q14, 2010 HSC)

12 Which of the following alternatives lists ALL of the chemical components of a chromosome?

(A) Sugar, phosphate and bases

(B) Lipids, DNA and protein

(C) DNA and protein

(D) Genes and DNA

(q4, 2011 HSC)

13 The presence of freckles is a dominant characteristic. A child's mother has no freckles and its father is heterozygous for freckles.

What is the probability that this child will have freckles?

(A) 25%

(B) 50%

(C) 75%

(D) 100%

(q5, 2011 HSC)

14 Which of the following correctly identifies a source of variation in both asexual and sexual reproduction?

	Asexual	*Sexual*
(A)	Cloning	Natural selection
(B)	Crossing over	Cell division
(C)	Spontaneous generation	Fertilisation
(D)	Mutation	Mutation

(q10, 2011 HSC)

15 Identical twins have the same genotype.

Why are there small differences between the phenotypes of identical twins?

(A) Some genes are not co-dominant.

(B) The environment affects the expression of genes.

(C) Both parents are homozygous for those phenotypes.

(D) Chromosomes segregate independently during meiosis.

(q8, 2012 HSC)

Part B Short-answer questions

Question 16 (4 marks) **Marks**

The diagram shows a cell containing three pairs of chromosomes just prior to a *meiotic* division.

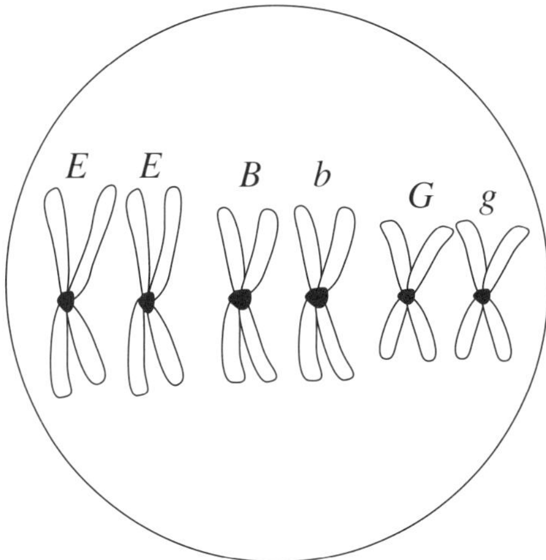

Assuming that random segregation occurs, construct a table that lists the possible genotypes that could be produced and states the expected frequency for each genotype. **4**

(5–6 line spaces) (q22, 2002 HSC)

Question 17 (6 marks) **Marks**

Examine the data collected on *Homo neanderthalensis* fossils found in recent years.

Date	*Data*
2008	• Full sequence of mitochondrial DNA of a range of Neanderthal fossils from within and between different fossil sites • The sequence is nearly identical within one fossil site • The sequence is very different between fossil sites
2010	• Full sequence of Neanderthal nuclear DNA • 1%–4% of genes in European modern humans are specific Neanderthal genes • No identifiable specific Neanderthal genes in modern sub-Saharan African humans • No specific modern human genes in nuclear DNA of Neanderthal fossils

(a) What inferences can be made about Neanderthal populations, based on the data collected in 2008? **3**

(6 lines)

(b) What inferences can be made about migration and breeding, based on the data collected in 2010? **3**

(6 lines) (q34(b), 2012 HSC)

Question 18 (5 marks)

Scientists have tried to achieve a viable embryo by fusing two ova (eggs) from the same female.

Explain whether the offspring produced using this process would be a clone of the female whose two ova were used. Use your knowledge of gamete formation and sexual reproduction to support your answer. **5**

(16 lines) (q29, 2014 HSC)

Question 19 (6 marks) **Marks**

(a) Cloning is a technique that could be used to increase numbers in an endangered species. What effect would cloning have on the genetic diversity of the species? **2**

(4 lines)

(b) Explain TWO possible evolutionary effects of a disease entering an endangered population containing some cloned individuals. **4**

(8 lines) (q22, 2001 HSC)

Question 20 (4 marks)

(a) On the diagram, clearly identify ONE nucleotide by placing a box around it. **1**

Structure of portion of a DNA molecule

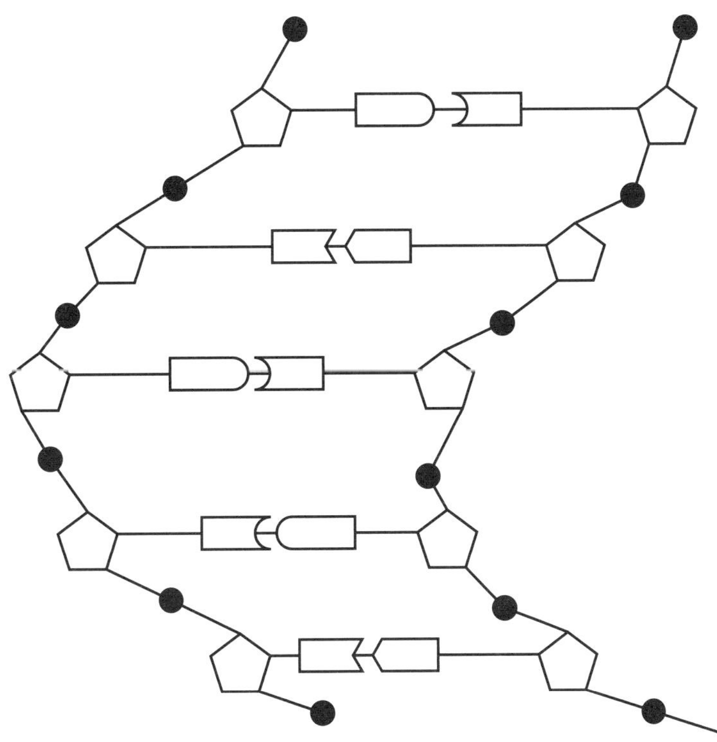

(b) Outline the main steps of DNA replication. **3**

(6 lines) (q16, 2004 HSC)

Question 21 (5 marks) **Marks**

(a) How could a mutation in DNA affect polypeptide production? 3

(6 lines)

(b) How could a change in a polypeptide affect cell activity? 2

(4 lines) (q25, 2008 HSC)

Question 22 (4 marks)

This flowchart represents a model of polypeptide production.

In the table below it, name and outline what occurs in Processes 1 and 2.

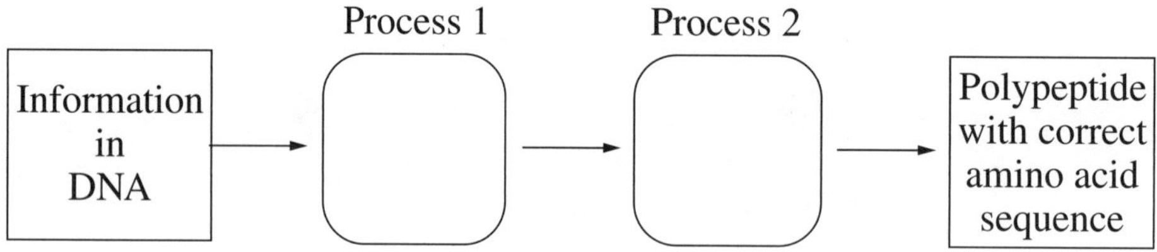

In the table below, name and outline what occurs in Processes 1 and 2. 4

Name of process (word or phrase)	*Outline of process*
Process 1	....
Process 2	....

(q18, 2005 HSC)

Question 23 (6 marks) **Marks**

Nick's wife Maria has a history of red–green colour blindness in her family. Jack, their two-year old son, may be red–green colour blind. Maria's brothers Vincent and Paul are colour blind but her brother James is not. Maria's mother Anne is a carrier of red–green colour blindness. Her father John is unaffected.

(a) Construct a family pedigree to show the inheritance of this sex-linked genetic disorder. **3**

(12 line spaces)

(b) Predict whether Jack will be colour blind. Justify your answer. **2**

(6 lines) (q24, 2005 HSC)

Question 24 (8 marks)

The diagram represents the DNA profiles from a murder case.

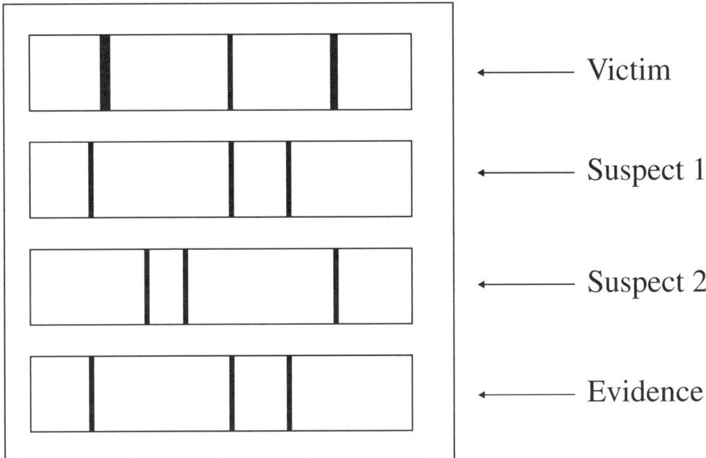

(a) Justify which one of the suspects could have been the murderer. **2**

(3 lines)

(b) Explain why a sample would be taken from the victim as well as the suspects. **3**

(4 lines)

(c) Outline the use of DNA fingerprinting as an investigation tool in forensic cases. **3**

(6 lines) (q29(d)(i), (ii) and (iii), 2005 HSC)

Question 25 (7 marks) **Marks**

Consider the following extract and the discussion following it.

> **DNA kits can bring unwanted surprises**
>
> For the price of a night out, individuals can learn key elements of their genetic composition and take treatment, or protect their children from hereditary health risks ...
>
> To take the test, a client spits into a test tube or swabs the inside of their cheek, then sends the sample for analysis ...
>
> Mrs X took the test to determine the cause of some minor health problems ...
>
> She learnt that she was at risk of breast cancer but also that the man she has called 'Dad' for 50 years was not her father. She tracked down her biological father and a half-sister, who had breast cancer. A biopsy found Mrs X had cancer.'
>
> (Source: Extract from Philip Sherwell, *The Telegraph*, London 7 January 2013)

Mutations in two genes BRCA1 and BRCA2 are particularly frequent in women with close relatives who suffer from breast, ovarian and other cancers or have early onset breast cancer. The risk of contracting breast cancer in an individual experiencing a mutation in these genes varies between 50% and 85%. There is a lack of general population data that prevents accurate prediction in individual cases.

(a) Outline the use of DNA sequencing and DNA profiling in determining the genetic information revealed to Mrs X. **2**

(6 lines)

(b) During your course you investigated the use of population genetics studies to determine the inheritance of a disease or disorder. Compare the results of your investigation with the use of genetic information to assess the risk of breast cancer. **5**

(10 lines) (q33(e), 2013 HSC)

Question 26 (6 marks)

Assess the effect of cell replication processes on continuity of the species. **6**

(12 lines) (Sample question)

Question 27 (7 marks) **Marks**

How do the features of fertilisation, implantation and the hormonal control of pregnancy and birth contribute to the continuity of species of mammals? **7**

(26 lines) (Sample question)

Question 28 (6 marks)

Outline the advantages of internal and external fertilisation in ensuring the continuity of species. **6**

(12 lines) (Sample question)

Question 29 (8 marks)

Evaluate the impact of scientific knowledge on the manipulation of plant and animal reproduction in agriculture. **8**

(26 lines) (Sample question)

Question 30 (4 marks)

Compare the impact of asexual reproduction with that of sexual reproduction on the continuity of plant species. **4**

(8 lines) (Sample question)

Question 31 (5 marks)

Using a specific example, explain how single nucleotide polymorphism is used to compare frequencies of characteristics in a population. **5**

(10 lines) (Sample question)

Question 32 (4 marks)

What is the role of genes and environment on phenotype? **4**

(8 lines) (Sample question)

Question 33 (6 marks)

Evaluate the impact of a population genetics study on the understanding of human evolution. **6**

(12 lines) (Sample question)

Question 34 (5 marks) **Marks**

Compare the role of DNA sequencing with DNA profiling in determining the inheritance patterns of a population. 4

(10 lines) (Sample question)

Question 35 (6 marks)

The following table provides information from a variety of sources about a major disease threat to the population of Tasmanian Devils, *Sarcophilus harrissi.*

Date	*Finding*
1996	A disease called the Devil Facial Tumour Disease (DFTD) was photographed in a wild Tasmanian Devil.
2008	This infectious, cancerous disease spread quickly from east to west across Tasmania so that the Devil population dwindled by about 70%. Concerns were expressed about the low rate of genetic diversity in the Devils, adding to their vulnerability to the disease.
2009	Tasmanian Devil were listed as endangered.
2011	The Devil Ark conservation breeding program began captive breeding in a facility containing 44 healthy Devils in Barrington Tops, New South Wales.
2015	The wild population of Tasmanian Devils reportedly declined by 90%. Twenty-two of Devil Ark's captively bred Devils were released into the wild in Tasmania.
2028	CSIRO scientists predict that by 2028 the Tasmanian Devil could be extinct in the wild.

(Compiled from:

Australian Government Department of Environment and Energy. (n.d.). Species Profile and Threats. Database at www.environment.gov.au/cgi-bin/sprat/public/publicspecies.pl?taxon_id=299.

Lunney, D., M. Jones and H. McCullum. (2008). Lessons from the looming extinction of the Tasmanian Devil. *Pacific Conservation Biology*, 14(3):151–153. Viewed online at https://doi.org/10.1071/PC080151.

Devil Ark animal preservation project, Barrington, New South Wales. Viewed online November 2018 at www.devilark.org.au)

Use this information and/or the results of your own investigations to explain the role of population genetics in conservation management. 6

(15 lines) (Sample question)

Module 5
Heredity

Answers

Part A Objective-response questions

1 D Each species is tolerant of a narrow temperature range, outside of which the species cannot exist.

2 A These are the only possible combinations from this crossing-over point.

3 A One of the parents had to be heterozygous for height if 50% of the offspring were short and one of the parents was homozygous recessive.

4 C The heterozygous rabbit is represented by Bb and a white rabbit would be homozygous bb.

5 B All other options involve artificial manipulation techniques.

6 A This represents a change in the sequence of bases and hence a mutation.

7 C New alleles are created by mutations whereas crossing over creates new combinations of alleles and hence variation.

8 A If 30% of bases are adenine, then another 30% are thymine, leaving 40% to be equally divided between guanine and cytosine, resulting in 20% each, respectively.

9 B A gamete has half the number of chromosomes of each parent so that when male and female gametes unite in fertilisation a full complement is restored.

10 B This is the only correct sequence indicating unzipping and complementary base pairing, resulting in identical copies of DNA.

11 D Non–sex-linked, because only one male offspring in the first generation possesses the characteristic. Dominant, because one male offspring in the second generation does not possess the characteristic even though both his parents are affected.

12 C Chromosomes are made up of 60% protein and 40% DNA.

13 B The mother's genotype is homozygous recessive and the father's is heterozygous for freckles.

14 D Cell division, spontaneous generation and cloning do not result in variation.

15 B While identical twins may be identical genetically, they may express these genes differently (phenotype) due to the environment in which they grow up.

Part B Short-answer questions

Question 16 (Total 4 marks)

Genotypes possible	*Frequency (%)*
EBG	25
EBg	25
EbG	25
Ebg	25

(4 marks)

Question 17 (Total 6 marks)

(a) The identification of a full sequence of mitochondrial DNA being identical in one site only suggests that the individuals were descended from a single matriarchal female because the mitochondrial DNA can only ever be inherited from the female and not the male. The females did not tend to move between sites as the mitochondrial DNA is very different between the sites. *(3 marks)*

(b) The data collected in 2010 suggests that European modern humans are descended from Neanderthals because 1% to 4% of genes are present in the population. As there are no specific Neanderthal genes in the modern sub-Saharan Africans, it is fair to say that the Neanderthals did not migrate outside of Europe. Without the presence of any specific modern human genes in the nuclear DNA of Neanderthal fossils, it could be said that ancestral species of the two did not interbreed but were separate lines. *(3 marks)*

Question 18 (Total 5 marks)

Ova are produced through the process of meiosis, which results in four daughter cells. Each daughter cell carries a haploid number of chromosomes. The chromosomes may be slightly altered by crossing over, independent assortment, and random segregation of alleles. These processes all result in possible genetic variation during meiosis, so the daughter cells will not be identical. Fusing two separate ova from the same female might create a full complement of chromosomes, but each will be slightly different, resulting in a unique offspring that is not genetically identical to the original female (i.e. different genotype). As such, the offspring will not be a clone of the original female, but will be female. *(5 marks)*

Question 19 (Total 6 marks)

(a) Cloning results in genetically identical offspring. This can reduce the frequency of diversity within a species' gene pool. *(2 marks)*

(b) Example: A disease entering a population of endangered organisms in which there are some cloned individuals and some normal individuals may have no effect at all if all of the individuals (the species and its clones) are resistant to it. However, if only the cloned individuals have resistance to this disease, then they would increase in the population through their greater survival and reproduction while the non-resistant individuals would die off. While disease resistance would be selected for, there would be a decrease in genetic diversity because only the cloned individuals having identical genotypes to the parent would survive. *(4 marks)*

Question 20 (Total 4 marks)

(a)

Structure of portion of a DNA molecule

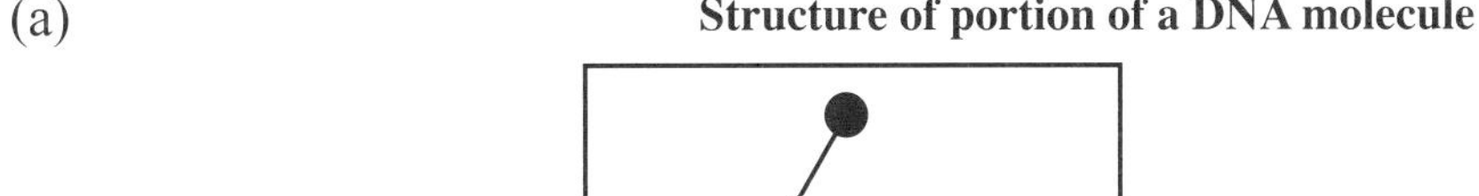

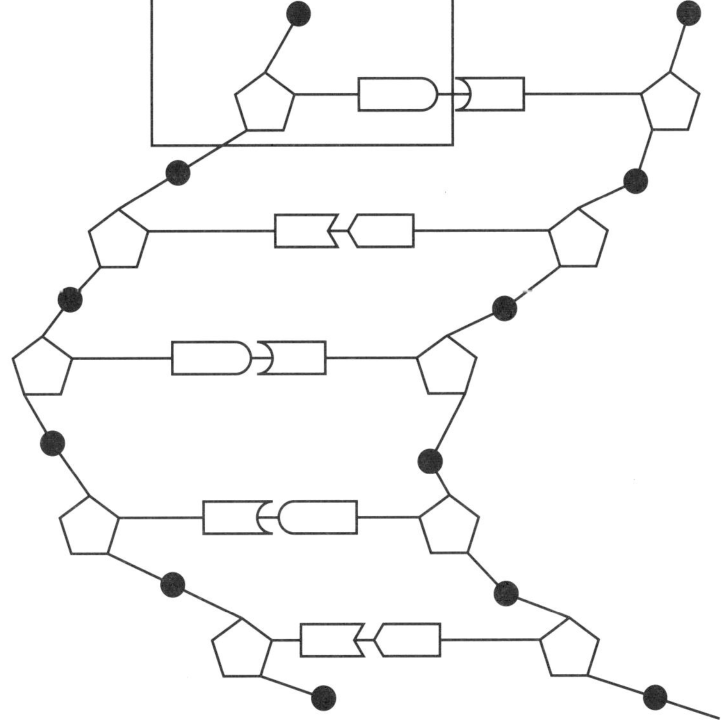

(1 mark)

(b) The double-stranded DNA unzips as the hydrogen bonds break. The resulting exposed bases pair up with free complementary nucleotides, giving rise to identical copies of the original DNA strand. *(3 marks)*

Question 21 (Total 5 marks)

(a) A mutation may vary in its impact, from causing no change in the polypeptide produced (i.e. a silent mutation), to one that results in a defective or no polypeptide being produced. A mutation in 'coding' DNA could directly result in an altered polypeptide. DNA point mutations can affect polypeptide synthesis because there is a substitution of nitrogenous bases, resulting in a different amino acid being included in the polypeptide. An insertion or deletion of a nitrogenous base in DNA causes a frameshift so that all the polypeptide synthesis 'downstream' of the mutation is affected.

DNA mutation in the 'non-coding' section of DNA may have varied impacts on polypeptide production. If the 'non-coding' mutation occurs in the part of DNA involved in controlling polypeptide synthesis, a polypeptide may be not be produced that would have normally been made, or it could be produced at different times or in different quantities. *(3 marks)*

(b) Polypeptides make up proteins that are one of the main structural and functional units of cells. A change in a polypeptide can result in the cell no longer being able to carry out its role. A change in a polypeptide that makes up the protein haemoglobin could result in red blood cells that are unable to carry sufficient oxygen to meet the needs of all cells. *(2 marks)*

Question 22 (Total 4 marks)

Name of process (word or phrase)	*Outline of process*	
Process 1 Unzipping/transcription	Double strand of DNA unwinds. RNA moves along strand linking complementary nucleotides together to form mRNA. The mRNA then moves from the nucleus into the cytoplasm.	*(2 marks)*
Process 2 Translation (polypeptide synthesis)	mRNA strand binds onto ribosome. tRNA binds to mRNA within the ribosome. The ribosome moves along the strand linking the amino acids, forming a polypeptide chain. When a stop codon is reached, the chain is released into the cytoplasm.	*(2 marks)*

Question 23 (Total 6 marks)

(a)

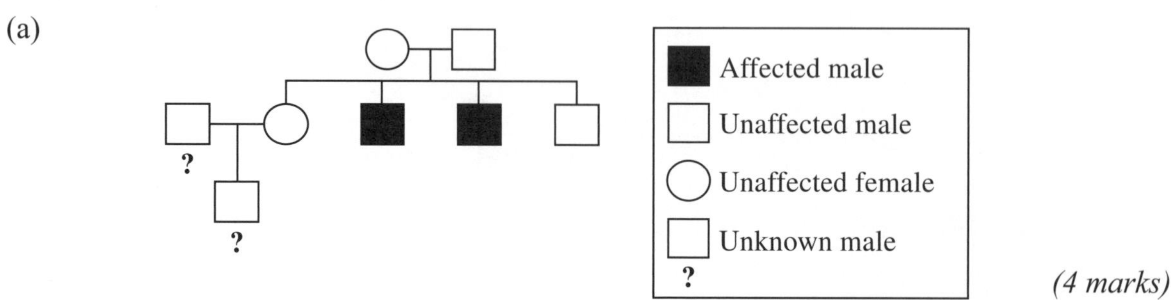

(4 marks)

(b) Jack may or may not be colour blind. Maria has 50% chance of being a carrier as her mother was a carrier; hence she has two colour-blind brothers. Jack has a 50% chance of being colour blind, depending on whether Maria is a carrier or not. *(2 marks)*

Question 24 (Total 8 marks)

(a) The DNA profile found as evidence is identical to the DNA profile of Suspect 1, which justifies the identification of Suspect 1 as the murderer. *(2 marks)*

(b) The DNA of both the victim and the suspects needs to be sampled in order to discount the victim as a suspect, to place either Suspect 1 or 2 at the crime scene, or to remove a suspect from the investigation. Samples taken at a crime scene can be easily contaminated by people coming into contact with the area (e.g. eyelash, skin cells), therefore the known DNA must be identified so that it can be excluded from the investigation. *(3 marks)*

(c) DNA fingerprinting involves extracting DNA from cells (e.g. blood, semen, hair roots), using restriction enzymes to cut the DNA into fragments that are highly specific for every individual. The fragments are made up of highly repetitive sequences of bases (variable number tandem repeats: VNTR) and vary in the number of bases and therefore size. These vary markedly between individuals and are inherited. A genetic fingerprint of dark and light bands is then produced for each sample by sorting these fragments using gel electrophoresis and radioactive probes. In unrelated people only one in four bands match by chance, so if two samples produce the same pattern, it is assumed that they come from the same person.

Comparisons of DNA fingerprints can investigate the likelihood of a relationship existing between the samples; for example, putting the suspect at the scene of the crime. *(3 marks)*

Question 25 (Total 7 marks)

(a) DNA sequencing determines the sequence of nitrogenous bases on all the DNA (i.e. the genome). Comparison of DNA sequences determined in the Human Genome Project has helped to identify genes responsible for some diseases such as genes associated with a high risk of breast cancer (BRAC1 and BRAC2). DNA profiling would compare only a small proportion of the DNA, the highly variable repeating sections of non-coding DNA that are used to confirm paternity or other familial relationships. *(2 marks)*

(b) (Answers will vary depending on the example chosen.)

Population genetics studies have recently confirmed the reason for the high incidence of the HbS allele in Africa. Abnormally long, red blood cells (sickle cells) is a condition associated with a single nucleotide polymorphism for haemoglobin, with the allele form HbS occurring in homozygous form in sufferers of sickle cell anaemia and in heterozygous form in those with the sickle cell trait. HbS allele frequency is particularly high in Africa where malaria is endemic. While sickle cell anaemia is often fatal, the trait form of the disease does not have serious implications in terms of blood flow and adds the benefit of providing resistance to the mechanism of action of the protozoan that causes malaria. *(3 marks)*

Identification of the BRCA1 and BRCA2 genes that increase the risk of breast cancer stem from studies of population genetics. In this case there appears to be no benefit of carrying mutations in genes that increase the risk of breast cancer but the identification of carriers of the genes can allow them to take preventative surgery. *(2 marks)*

Question 26 (Total 6 marks)

In eukaryotes the two main cell replication processes include mitosis and meiosis. Both of these processes follow a 'resting' or interphase part of the cell cycle. One of the earliest processes is duplication of DNA, followed by condensation of the nuclear material, chromatin, to form chromosomes. The chromosomes consist of two chromatids that are attached at a centromere.

MITOSIS: The separation of these chromatids, one to each daughter cell, results in the faithful replication of the genetic material in mitosis. Mitosis assists the continuity of species by allowing growth and repair as well as asexual reproduction. It contributes little to genetic diversity but greatly to population size.

MEIOSIS: During meiosis, the cell division essential for sexual reproduction— the chromatids of adjacent homologous chromosomes—may undergo crossing over to result in a new assortment of alleles. Meiosis is a two-stage cell division that results in half the chromosomes in each daughter. This is the result of the random segregation of the homologous pairs. Both the crossing over and the random segregation are the source of genetic variation which, once combined with the fertilisation from another member of the species, means that sexual reproduction results in new members of the species that are genetically varied.

In prokaryotes and some eukaryotes, cell replication processes may include binary fission, budding and spore production. The structure and arrangement of the DNA in prokaryotes is more primitive so replication is often simple and rapid, helping to ensure large numbers and species' continuity. *(6 marks)*

(2 marks to identify key processes of cell replication that ensure genetic continuity and 4 marks for outlining how the processes in mitosis, meiosis and binary fission assure species' continuity through growth, repair and reproduction)

Question 27 (Total 7 marks)

Fertilisation is essential for sexual reproduction of all mammals. Not only does it result in the formation of a zygote but also, because it involves the genetic input of two different parents that produced haploid gametes, it promotes genetic diversity, which is important for natural selection and evolution in changing environments.

Implantation is essential for placental mammals and briefly for marsupials whose embryos develop temporarily in a uterus. Monotremes that lay eggs do not rely on implantation for continuity of the species. Implantation allows the internal development of the embryo (and foetus in the case of placental mammals) so that young are born in a relatively developed state.

The hormonal control of pregnancy and birth in mammals ensures that the foetus is born when it has developed sufficiently to allow its survival. Progesterone ensures that the uterus is prepared and able to support the development of the placental and provide nourishment for the foetus. Oestrogens strengthen the musculature of the uterus. Oxytocin helps initiate muscular contractions of the uterus to bring about dilation of the cervix and birth. Oxytocin also supports maternal bonding with the offspring. Relaxin allows the expansion of joints and muscles of the mother to allow the passage of the foetus in birth. Corticotrophin-releasing hormone is believed to trigger birth when its levels dramatically increase.

The combined impact of these and other hormones is the maintenance of pregnancy and the timing of birth to ensure the successful delivery of offspring and hence the continuity of a species. *(7 marks)*

(2 marks each for describing the key features of fertilisation, implantation and the hormonal control, and 1 mark for relating each to the continuity of species)

Question 28 (Total 6 marks)

Internal fertilisation increases the survival rate of the male and female gametes, and increases the chances of fertilisation. Fluids in semen that help protect and nourish the spermatozoa are retained. Internal fertilisation is associated with a relatively small number of oocytes being produced and therefore is less physically demanding of the female. Internal fertilisation allows the internal development of the embryo and foetus so that they are better protected from environmental changes and predators. Internal fertilisation allows the development of a shell around the egg (soft in reptiles, hard in birds) that protects the embryo and its nutrition until hatching.

External fertilisation is often associated with aquatic species and it allows dispersal of the embryos. External fertilisation is associated with a high production of gametes in both males and females but then requires little parental energy to raise the offspring. In the case of seahorses, external fertilisation has allowed the males to house and protect the developing eggs in a pouch to increase survival rates. Some argue that this frees the female to commence egg production sooner. *(6 marks)*

(3 marks each for outlining advantages of internal and external fertilisation)

Question 29 (Total 8 marks)

An understanding of asexual reproduction in plants has been used to propagate large quantities of fruit trees and some vegetables. Effective cloning, propagating from bulbs, corms, tubers and runners, and cutting and grafting produces many plants with predictable qualities. Tissue culture using aseptic techniques is used also to produce large quantities of disease-free plants such as bananas.

An understanding of sexual reproduction in plants, including knowledge of flower structure and pollination methods, has been used to artificially pollinate a large number of plant varieties to increase yield and produce hybrids with improved qualities. Hybrid vigour is the result of hybridising two varieties of plants, with the resulting hybrid having qualities or yields that supersede either of the parent varieties. Triticale is a hybrid of wheat and rye that is purported to have yields and nutritional benefits beyond wheat and rye.

An understanding of chromosome arrangements and numbers has been used to produce some sterile varieties (e.g. triploid) of plants, such as seedless watermelons. Polyploidy (having more sets of chromosomes than the normal diploid set) is linked with increased cell and plant size, and productivity of many plants in agriculture.

An understanding of genetics has supported the success of artificial selection in breeding better plants (and animals) for agriculture. Mutagens have been used to increase the diversity of plants that then are selected if they have improved horticultural potential.

An understanding of sexual reproduction in animals, including fertilisation and the hormonal control of ovulation, implantation, pregnancy and birth, has been used to create procedures such as artificial insemination, in-vitro fertilisation and embryo production. Oestrous synchronisation of female cattle increases the efficiency of artificial insemination. Multiple ovulation and embryo flushing and transfer can increase the reproductive potential of high-quality female animals. Selecting sex is beneficial in breeding dairy animals (where female offspring are preferred) and aquaculture. Deliberately producing sterile triploid fish can improve productivity or prevent the interbreeding of genetically modified organisms with wild populations.

Application of understanding of plant and animal reproduction from the beginning of selective breeding to modern genetic technologies has had vast implications for the improvement of agricultural productivity, resistance to disease and improved quality of produce. *(8 marks)*

(3 marks each for explaining plant and animal agricultural improvements, 1 mark for genetic improvements resulting from scientific understanding and 1 for linking them to agricultural benefits)

Question 30 (Total 4 marks)

Population size and diversity are the keys to plant species' continuity. Plants also often rely on reproduction to achieve dispersal, which is also a key to continuity.

Asexual reproduction contributes greatly to population size and short-range dispersal. Structures that store food such as bulbs, rhizomes and corms may contribute to continuity by assisting plants to reproduce after changes in seasons or environmental changes.

Sexual reproduction in plants contributes to genetic diversity and significant and more widespread species dispersal. Pollination methods vary, with wind pollination contributing significantly to gymnosperms and many of the grass angiosperms. Wind pollination requires large quantities of pollen to ensure pollination and fertilisation but supports dispersal. Self-pollination ensures seed production but at the expense of less diversity. Insect, bird and mammal pollinators allow diversity, dispersal and population expansion but occur at the expense of interdependence on pollinator species. Seed size and structure contribute to the successful dispersal of both gymnosperms and angiosperms. *(4 marks)*

(2 marks for explaining how reproductive features impact on plant species' continuity and asexual reproduction in particular and 2 marks explaining how sexual reproduction contributes to species' continuity in plants)

Question 31 (Total 5 marks)

A single nucleotide polymorphism (SNP) is a result of a mutation resulting from the substitution of a nucleotide base that has occurred and is maintained in at least 1% of the population. They are the most common form of genetic variation in the human population. They occur in all parts of the DNA: the coding regions and the regulatory and non-regulatory regions of non-coding DNA and so may have varying impacts or no impact on health or development.

Sickle cell disease is a result of a SNP in the coding region for the beta globin polypeptide that makes up haemoglobin. If it is homozygous, the condition is the serious disease(sickle cell anaemia) but if heterozygous, the sickle cell trait has only minimal deleterious impact on circulation and appears to promote resistance to malaria. The incidence of the SNP correlates highly with the risk of contracting malaria; that is, it is frequent in sub-Saharan Africa and parts of India. *(5 marks)*

(3 marks for explaining the use of SNP to compare frequencies of characteristics in a population and 2 marks for describing a specific example)

Question 32 (Total 4 marks)

Genes code for the production of polypeptides that make up proteins. Proteins have structural and functional roles in cells and organisms and so contribute significantly to the phenotype of organisms.

The environment of an organism—external, internal and even in the womb—can play a role in the expression or regulation of genes and so also impact on the phenotype. Epigenetics is the study of how the environment can impact on an organism and even the offspring. Studies of women in their first trimester of pregnancy in the Dutch Hunger Winter resulting from World War II showed that, even if their offspring were born in the normal weight range (because of improvement in food intake later in the pregnancy), they suffered abnormally high health problems such as obesity and mental health problems later in life. *(4 marks)*

(2 marks for explaining how genes impact phenotype and 2 marks for explaining how the environment impacts and interacts with genes)

Question 33 (Total 6 marks)

Population genetics attempts to model the dynamics of evolutionary change within and between populations. Mutation is the ultimate source of genetic diversity. The diverse gene frequency is the result of natural selection, genetic drift and gene flow. Random mating can influence the frequency of alleles in small populations (genetic drift) and migration tends to reduce differences across population groups (gene flow). Population genetics uses data on the diversity of modern humans but has also used DNA extracted from a small number of Neanderthal fossil finds.

If divergent evolution was a key factor in human evolution, it could be predicted that human 'races' are a reality reflected in genetic variation. It would be as if the various 'racial' groups were on a path towards evolution of separate species. If races exist there should be greater diversity across human population groups than within. This is not the case. Gradients in diversity rather than distinct differences based on geography have dispelled the notion of racial groups. The conclusion to be drawn is simply that anatomically and genetically humans are diverse.

The fossil record suggests Africa is the place where modern humans first evolved. The 'out of Africa' theory and 'theory of regional continuity' suggest human migration occurred in either two or one major wave from Africa. Gene flow is believed to have prevented evolution into separate species (in the theory of regional continuity). Genetic evidence now suggests ancient contributions from southern Asia. Some researchers now claim that human origins are best described as 'mostly (but not exclusively) out of Africa'.

Population genetic studies is now playing an increasing and vital role in understanding human evolution. *(6 marks)*

(2 marks for outlining the role of population genetics in investigating human evolution and 4 marks for evaluating ways it has contributed to understanding)

Question 34 (Total 5 marks)

DNA sequencing involves determining the sequence of nitrogenous bases in the entire genome of organisms. The Human Genome Project completed in 2003 used sample humans to determine a 'baseline' of the entire sequence in humans. From this it is possible to detect single nucleotide polymorphisms to get a measure of human diversity. The sequence can be used to estimate risks of a range of genetically linked diseases and the level of diversity across population groups and even species. DNA sequencing of nuclear, mitochondrial and Y chromosome DNA has been used to develop better understanding of human evolution and migratory history.

DNA profiling provides information about comparisons of DNA samples. It selectively compares parts of the 'non-coding' DNA that in short sequences repeats itself a variable number of times between individuals (short-tandem repeats or variable-number tandem repeats). It can be used to match DNA from crime scenes with possible suspects or compare individuals with possible familial relationships. It can be used to identify human remains if DNA profiles from relatives are available.

DNA sequencing is useful in determining inheritance patterns on a global level, whereas DNA profiling is useful at a family level. *(5 marks)*

(2 marks for outlining the role of DNA sequencing and 2 marks for DNA profiling; 1 mark for comparing the role of sequencing with profiling)

Question 35 (Total 6 marks)

Endangered species such as Tasmanian Devils or Wollemi Pine are characterised by small populations, fragmented or localised distribution patterns and low genetic diversity. They are susceptible to disease (as is the case with the Tasmanian Devil), habitat destruction and environmental change.

Population genetics helps understand and minimise the impact of low genetic diversity. DNA sequencing can be used to identify individuals that may be more diverse or disease-resistant and use them extensively in breeding. Constructing wildlife corridors can try to maximise opportunities for natural cross breeding to maintain genetic diversity. In the conservation management of plant species, finding opportunities to maximise breeding by sexual as opposed to asexual propagation should try to maintain genetic diversity.

Establishing captive breeding programs for Tasmanian Devils relies on maximising genetic diversity within each captive population and, from time to time, exchanging disease-free members in different populations.

In the conservation management of plant species such as the Victorian rice flower *Pimelea spinescens*, artificial pollination can be used to try to maintain diversity. Genetic markers (microsatellites) are being developed for *Pimelea spinescens.* These markers will provide further information on genetic variation across the species range. *(6 marks)*

(4 marks for explaining how population genetics contributes to understanding factors associated with threats to endangered species and 2 marks for identifying factors in the example provided or linking to a different investigation into conservation of an endangered species)

CHAPTER 2

Module 6
Genetic Change

Part A Objective-response questions

1 Which of the following is true of a mutation that produces an allele that is dominant?

(A) It would be expected to cause death.

(B) It would be expected to spread more quickly through a population than a recessive mutation.

(C) It could give an observable phenotype in a heterozygous genotype.

(D) It could give an observable phenotype only in a homozygous genotype.

(q7, 2001 HSC)

2 How can widespread use of artificial insemination alter the genetic composition of a population?

(A) It results in many genetically identical individuals.

(B) It makes certain alleles more common in a population.

(C) It decreases the number of chromosomes in some individuals.

(D) It ensures that only the genetic composition of the males is altered.

(q11, 2010 HSC)

3 The following events occur after DNA is subjected to radiation. The events are listed in no specific order.

P: Change in protein structure

Q: Change in polypeptide sequence

R: Change in cell activity

S: Mutation

What is the correct sequence of steps?

(A) S, P, Q, R

(B) S, Q, P, R

(C) R, Q, S, P

(D) R, S, Q, P

(q16, 2010 HSC)

4 What is the name of the process that results in organisms containing DNA from different species?

(A) Transcription

(B) Transgenics

(C) Translation

(D) Translocation

(q1, 2012 HSC)

5 Which of the following is an example of *hybridisation*?

(A) The insertion of a bacterial gene for herbicide resistance into a cotton plant.

(B) The culturing of a cell taken from the root of a carrot to form a small plant.

(C) Artificial insemination of a domestic cat with wild cat semen to produce a Bengal cat.

(D) A cutting taken from one variety of apple tree grafted onto the stem of a different variety of apple tree.

(q12, 2012 HSC)

6 Reproductive technologies focus on the transfer of genetic information.

Which process only involves the transfer of the nucleus?

(A) Cloning

(B) Transgenesis

(C) Artificial pollination

(D) Artificial insemination

(q12, 2004 HSC)

7 An experiment was conducted to examine the effect of ultraviolet radiation on the development of antibiotic resistance in a strain of bacteria. The table summarises the outcomes of this experiment.

Which of the following statements best summarises the stages in the development of the new strain of bacteria that was resistant to antibiotic *S*?

	Antibiotic resistance				
Treatment	*Antibiotic P*	*Antibiotic Q*	*Antibiotic R*	*Antibiotic S*	*Antibiotic T*
No exposure to ultraviolet radiation	✓	✓	✗	✗	✗
Exposure to ultraviolet radiation	✓	✓	✗	✓	✗

✓ = resistant ✗ = not resistant

(A) Hybridisation → Mutation → Natural Selection

(B) Replication → Mutation → Natural Selection

(C) Mutation → Natural Selection → Replication

(D) Mutation → Hybridisation → Natural Selection

(q5, 2002 HSC)

8 To protect a farm animal from a plant toxin, a gene for resistance to the toxin was transferred to the farm animal.

Which term best describes this process?

(A) Cloning

(B) Genetic engineering

(C) Artificial pollination

(D) Artificial insemination

(q3, 2008 HSC)

9 Consider the following image.

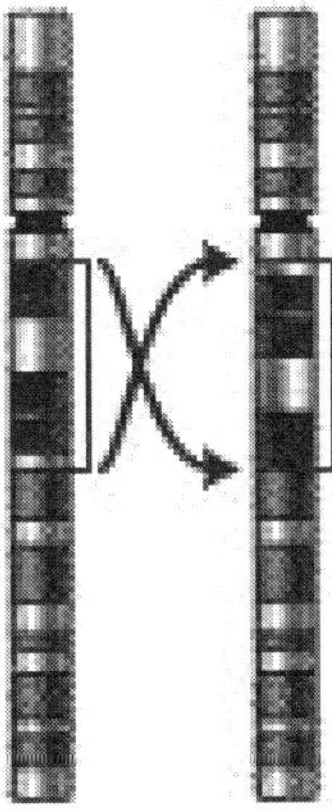

(Source: Courtesy National Human Genome Research Institute. (2006). Viewed November 2018 at https://commons.wikimedia.org/wiki/File:Mutation_inversion.jpg. Wikimedia Creative Commons/Public Domain CC-BY-SA-3.0)

What does the image represent?

(A) Substitution gene mutation

(B) Substitution chromosome mutation

(C) Frameshift gene mutation

(D) Inversion chromosome mutation

(Sample question)

10 The following table outlines two mutations.

	DNA Sequence
Original DNA	AAC TCG GTC AAT ATG
Mutation 1	AAC TCC GTC AAT ATG
Mutation 2	AAC TCG GTA ATA TGC

Which correctly row identifies the type of mutation that is mutation 1 and mutation 2?

	Mutation 1	*Mutation 2*
(A)	Substitution	Deletion
(B)	Substitution	Chromosome mutation
(C)	Insertion	Deletion
(D)	Deletion	Insertion

(Sample question)

Study the following diagram to answer Questions 11 and 12.

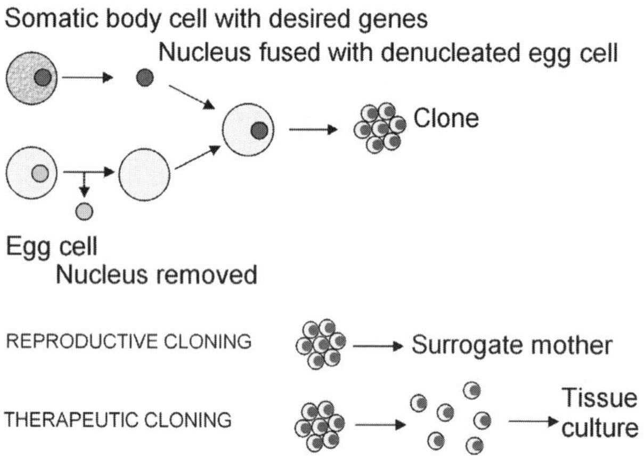

Somatic body cell with desired genes
Nucleus fused with denucleated egg cell
Clone
Egg cell
Nucleus removed
REPRODUCTIVE CLONING
Surrogate mother
THERAPEUTIC CLONING
Tissue culture

(Source: © Dr Jürgen Groth. (2005). Reproductive and therapeutic cloning diagram. Modified and with text translated by Wikibob. Viewed November 2018 at https://commons.wikimedia.org/wiki/File:Cloning_diagram_english.png. Wikimedia Creative Commons/CC-BY-SA-3.0/de:Vorlage:Bild-GFDL)

11 What is the best title for this diagram?

(A) Gene cloning

(B) Somatic-cell nuclear-transfer cloning

(C) Embryo-splitting whole organism cloning

(D) Artificial insemination cloning

(Sample question)

12 Which of the following statements correctly explains the purpose of one form of cloning outlined in the diagram?

(A) It produces transgenic plasmids that can make proteins such as insulin.

(B) It produces offspring that exhibit hybrid vigour.

(C) It produces cells that are used in gene therapy.

(D) It produces many offspring from high-quality parents.

(Sample question)

Read the following case study to answer Questions 13 and 14.

Hawaiian taro (kalo) is regarded as an important, almost sacred staple crop by the local Hawaiian community. The number of kalo varieties has declined by almost 50% in the last 100 years due to invasive pests such as pocket rot, with up to half the kalo crop being lost on an annual basis as a result of disease. The Solomon Islands has also recently had taro crops devastated by viral attacks. Despite the risks to the kalo production, the local Hawaiian community has presented an argument that has resulted in the University of Hawaii agreeing to suspension of research into genetic engineering of kalo.

Papaya is an export crop worth millions of dollars to Hawaiian farmers. Papaya was only relatively recently introduced to Hawaii. A new variety of papaya (Rainbow) has been genetically engineered to resist ringspot virus and it now accounts for well over half of Hawaiian papaya production.

13 What type of context influencing the acceptance of gene technologies on kalo is this case study an example of?

(A) Social

(B) Environmental

(C) Economic

(D) Cultural

(Sample question)

14 Which type of contexts most probably influenced the acceptance of genetic modification of papaya in Hawaii?

(A) Social and economic

(B) Ethical and environmental

(C) Cultural and social

(D) Ethical and economic

(Sample question)

15 What is the initial impact of growing transgenic crops on biodiversity?

(A) Increases biodiversity

(B) Decreases biodiversity

(C) Maintains biodiversity

(D) Decreases biodiversity locally but increases it globally

(Sample question)

Part B Short-answer questions

Marks

Question 16 (4 marks)

Consider the following survey summary.

> *Genetically modified food on menu*
>
> Australians are growing more accepting of genetically modified (GM) food, a federal survey has found.
>
> The survey found that close to 45 per cent of Australians now believe GM food would become more widely accepted and less risky in the next few years.
>
> An earlier survey revealed that people thought the risks outweighed the benefits, but that the situation would change.

(a) State ONE opinion held by Australians about genetically modified food, according to this article. **1**

(2 lines)

(b) Justify ONE piece of information you would need in order to determine the validity of such survey results. **3**

(6 lines) (q16, 2001 HSC)

Question 17 (6 marks)

(a) Cloning is a technique that could be used to increase numbers in an endangered species. What effect would cloning have on the genetic diversity of the species? **2**

(4 lines)

(b) Explain TWO possible evolutionary effects of a disease entering an endangered population containing some cloned individuals. **4**

(8 lines) (q22, 2001 HSC)

Question 18 (8 marks) **Marks**

Evaluate the impact of major advances in scientific understanding and technology, in the field of genetics, on developments in reproductive technologies. **8**

(26 lines) (q28, 2001 HSC)

Question 19 (4 marks)

With reference to TWO types of cloning, copy and complete the following table. **4**

Type of cloning	*Process used*	*Example*

(4 lines) (q33(a), 2012 HSC)

Question 20 (8 marks)

Explain how our knowledge of chromosome structure has led to reproductive technologies that have the potential to alter the path of evolution. 8

(26 lines) (q30, 2013 HSC)

Question 21 (4 marks)

Traditionally, banana plants in Australia have been propagated asexually by cutting out and planting suckers from the adult plant.

There is a growing trend to produce disease-free plants in laboratories through a process of cloning from disease-free tissues from existing plants.

Assess the potential impact of this cloning process on the genetic diversity of banana plants in Australia. **4**

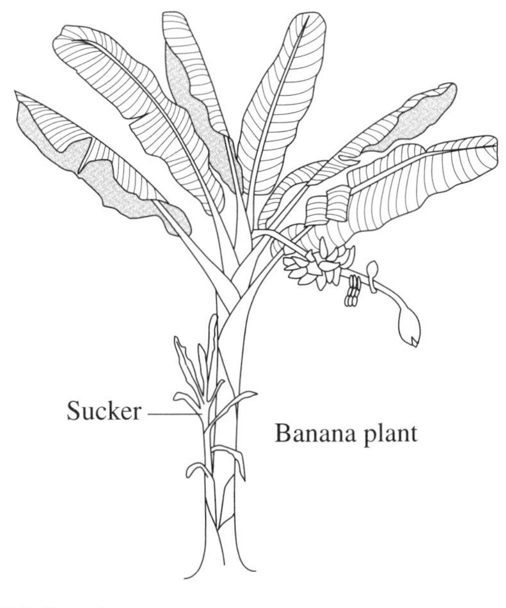

(11 lines) (q6, 2002 HSC)

Question 22 (3 marks) **Marks**

The diagram shows various forms of radiation that are part of the electromagnetic spectrum.

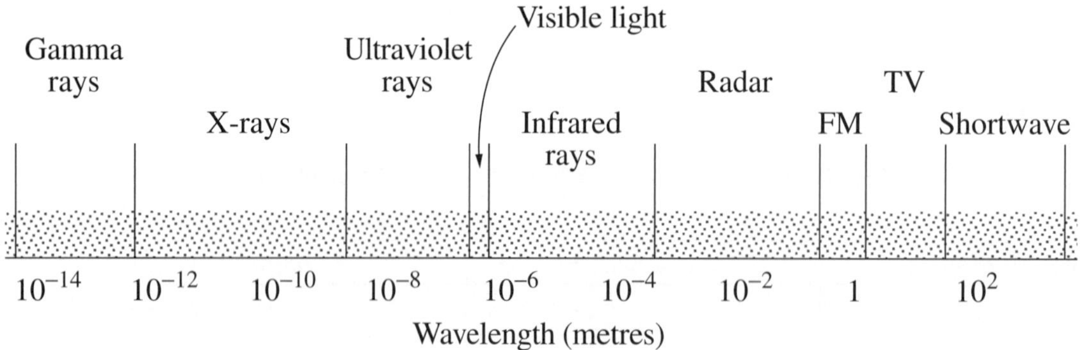

(a) Select ONE form of radiation that is considered to be a mutagen. **1**

(1 line)

(b) Describe evidence that supports the mutagenic nature of the selected form of radiation. **2**

(4 lines) (q20, 2003 HSC)

Question 23 (8 marks)

(a) Identify a transgenic species, and state its use. **2**

(3 lines)

(b) Outline ONE process used to produce a transgenic species. **2**

(4 lines)

(c) Discuss ONE ethical issue arising from the use of transgenic species. **4**

(7 lines) (q22, 2006 HSC)

Question 24 (3 marks)

'All mutations are harmful.' Discuss this statement. **3**

(6 lines) (q28, 2004 HSC)

Question 25 (8 marks) **Marks**

For this genetically modified fish to be produced, a number of processes would have been used. Gene splicing may have been one of them.

Fluorescent fish go on sale

Despite objections from food safety and conservation groups, zebra fish genetically modified to fluoresce red went on sale in several pet stores last week. 'They're selling really well,' says Steven Frenberg of the pet store chain. Each GloFish costs $4.99, ten times as much as a normal zebra fish, he says.

(a) Outline the process of gene splicing to produce recombinant DNA. **2**

(4 lines)

(b) Fluorescent zebra fish are able to mate with non-fluorescent fish to produce viable offspring. Propose an impact on the environment if the fluorescent fish are accidentally released. **2**

(4 lines)

(c) Explain why some groups in society may have different views about the use of DNA technology to produce zebra fish that fluoresce. **4**

(8 lines) (q31(d), HSC 2004)

Question 26 (3 marks) **Marks**

The flowchart shows the major steps in the production of recombinant DNA.

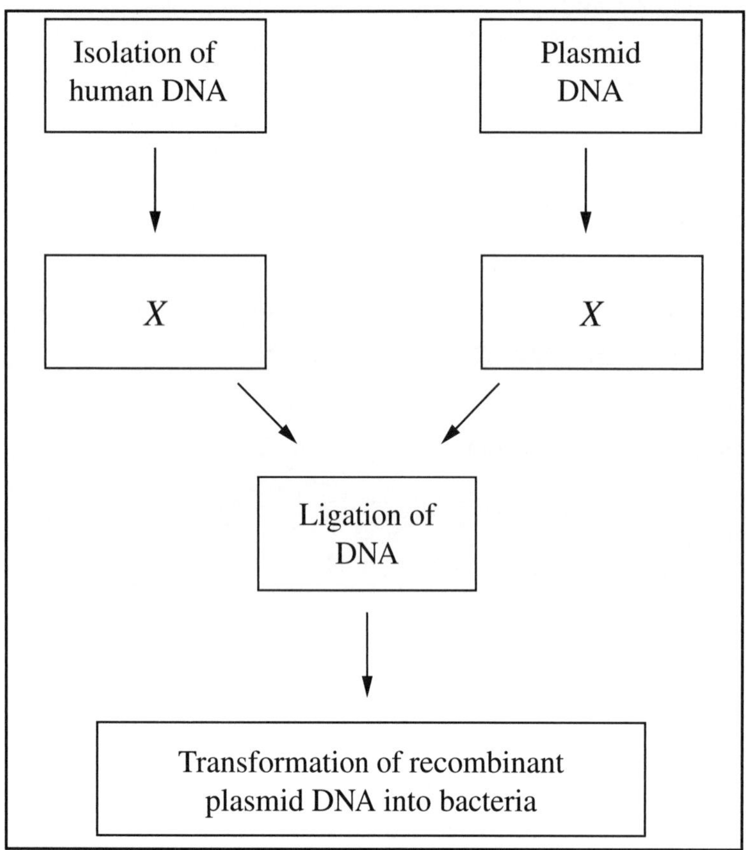

(a) Name the process labelled *X*. 1

(1 line)

(b) Outline the process of *ligation of DNA*. 2

(4 lines) (q30(a), 2001 HSC)

Question 27 (5 marks) **Marks**

The diagram illustrates three steps involved in the Polymerase Chain Reaction (PCR).

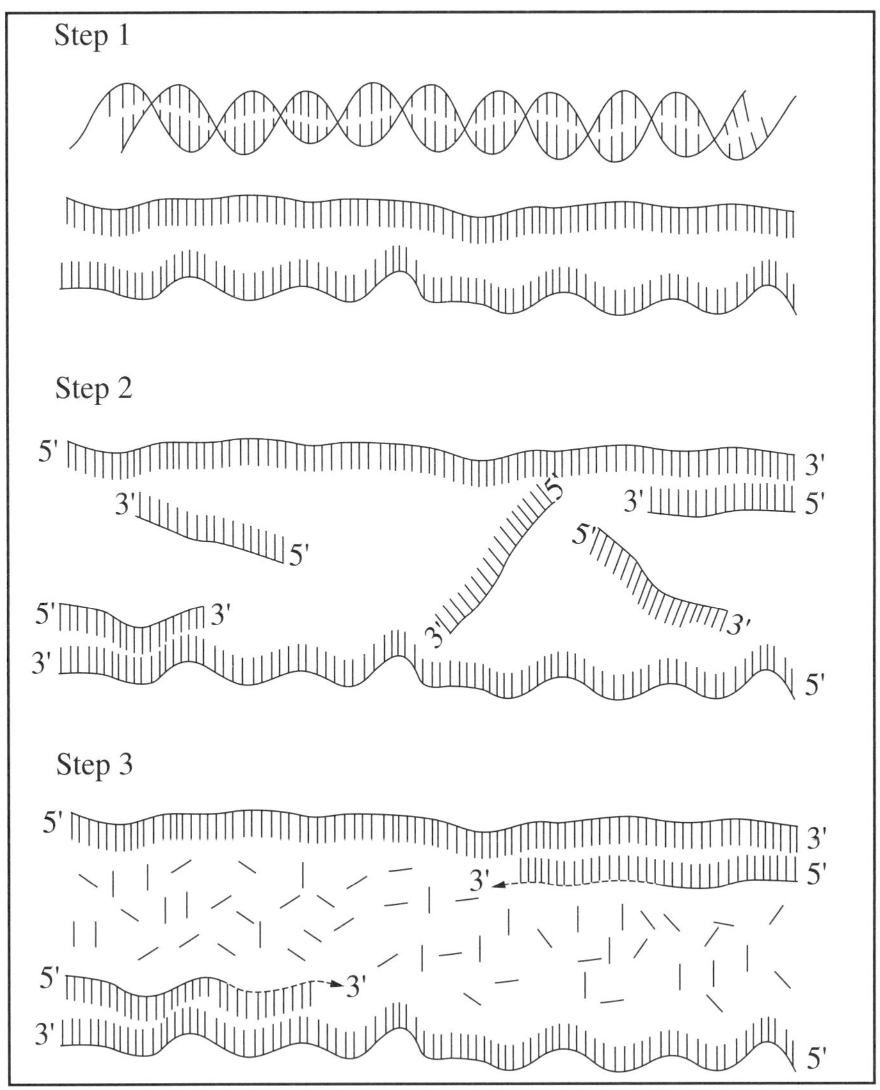

(a) State ONE difference between RNA and DNA. 1

(1 line)

(b) Give ONE use of PCR. 1

(2 lines)

(c) State what is happening in each of the steps shown in the above diagram. 3

(6 lines) (q29(a), 2002 HSC)

Question 28 (7 marks) **Marks**

Explain how applications of modern biotechnology have produced products that are useful to humans. 7

(27 lines) (q29(e), 2009 HSC)

Question 29 (6 marks)

Fire and Mello discovered a cellular process now known as RNA interference.

In this process, small nucleotides bind to specific mRNA sequences, causing their destruction. This discovery is revolutionising molecular therapeutics for devastating diseases.

(a) Explain how protein synthesis is affected by RNA interference. 2

(4 lines)

(b) Describe the roles of DNA and RNA in an application of biotechnology. 4

(8 lines) (q32(d), 2010 HSC)

Question 30 (4 marks)

Draw a flowchart showing the sequence of events that results in the formation of recombinant DNA. 4

(8 lines) (q32(b), 2010 HSC)

Question 31 (5 marks)

Consider the following passage.

> *Ants cultivate grass for food*
> A species of ant has been observed to clear the surface of the ground for 5 metres around its colony and bury grass seeds there. The ants remove any other plants that may spring up among their crop. When the seeds are ripe, the ants harvest the seeds for food and retain some seeds for next year's crop. A ten-year study has shown that the average size of the grass seeds increased by 5%.

(a) Outline the difference between qualitative and quantitative observations, using examples from the text above. 2

(4 lines)

(b) Explain why the ant activity could be interpreted as biotechnology. 3

(6 lines) (q33(c), 2011 HSC)

Question 32 (7 marks) **Marks**

The text below summarises some recent scientific experiments.

> Scientists studying the development of human female embryos recently discovered a gene called XIST. This gene silences one of the two X chromosomes so that they do not over-function in normal human females.
>
> The scientists were then able to insert the XIST gene into human cells grown in tissue culture to successfully silence other chromosomes.
>
> Scientists are now attempting to insert the XIST gene into the extra chromosome of mice that have trisomy.

With reference to genetics and gene technologies, explain these experiments and their implications. **7**

(26 lines) (q33(e), 2014 HSC)

Question 33 (8 marks)

Complete the following table. **8**

	Point mutations	*Chromosomal mutations*
Causes		
Processes		
Effects		

(26 lines) (Sample question)

Question 34 (6 marks)

Compare the action of chemical mutagens with naturally occurring mutagens. **6**

(26 lines) (Sample question)

Question 35 (5 marks)

Outline the procedures you undertook when investigating and assessing the effectiveness of whole organism cloning to ensure you extracted reliable information and drew accurate conclusions. **5**

(12 lines) (Sample question)

Module 6
Genetic Change

Answers

Part A Objective-response questions

1 C There is no reason to suggest that all mutations are lethal. All alleles have a 50% chance of being passed on and can produce an observable phenotype in both a heterozygous and homozygous genotype.

2 B Artificial insemination will make desired characteristics more prevalent in a population because greater numbers of stock will possess those characteristics and can pass them on.

3 B This is the only correct sequence indicating mutation and change as a result of radiation and consequently change to cell activity.

4 B Transgenic species are made by the insertion of another specie's genetic information. This is passed onto the next generation.

5 C The use of two varieties to produce unique characteristics of apple tree is an example of hybridisation. Under usual circumstances the act of mixing different species or varieties of animals or plants produces hybrids.

6 A Cloning involves the transfer of only the nucleus. The other processes rely on the transfer of whole cells.

7 C This is the only possible sequence because the bacteria had to become different first before resistance could develop and then be selected before passing on the new characteristics.

8 B Genetic engineering enables the alteration/modification of a genome by the insertion or deletion of genes.

9 D The diagram shows a section of chromosome being 'turned upside down' or inverted.

10 A Mutation 1 involves the substitution of a C (cytosine) for a G (guanine) nitrogenous base, whereas Mutation 2 is the deletion of the C (cytosine) base and subsequent frameshift of following bases.

11 B The key part of the procedure is the transfer of the nucleus from a somatic cell, whether it is used for whole organism cloning or therapeutic tissue culture.

12 C The function of therapeutic cloning is to produce cells that can be used to replace damaged or dysfunctional cells.

13 D The Hawaiian locals' regard of the kalo as sacred suggests that the influence to restrict research is based on cultural values.

14 A Resisting the virus would result in more profitable production of papaya and would impact positively, with economic and social benefits for farming communities.

15 A Initially the newly introduced genes into another species will increase the biodiversity.

Part B Short-answer questions

Question 16 (Total 4 marks)

(a) Australians now believe that genetically modified food will become more widely acceptable to people. *(1 mark)*

(b) (Answers could include information about the questions asked, survey sample size, structure and composition of the sample. An example is provided.)

The sample size used for the survey is important. The validity of the survey increases with a greater sample size, as this represents a closer opinion of the population being surveyed. The most valid surveys would include the whole population. *(3 marks)*

Question 17 (Total 6 marks)

(a) Cloning results in genetically identical offspring. This can reduce the frequency of diversity within a species' gene pool. *(2 marks)*

(b) Example: A disease entering a population of endangered organisms in which there are some cloned individuals and some normal individuals may have no effect at all, if all the individuals (the species and its clones) are resistant to it. However, if only the cloned individuals have resistance to this disease, then they would increase in the population through their greater survival and reproduction while the non-resistant individuals would die off. While disease resistance would be selected for, there would be a decrease in genetic diversity because only the cloned individuals having identical genotypes to the parent would survive. *(4 marks)*

Question 18 (Total 8 marks)

(The answer should include more than one advance in scientific understanding and technology. These should be clearly identified and evaluated in terms of their impact on developments in reproductive technology.)

Some advances include: identification of chromosomes and the mechanism of inheritance; DNA structure; mapping of the genome; and improved microscopes.

Some technological improvements in reproductive technology include: artificial insemination, cloning, sex selection, and transgenic species.

Example: Advances in tissue culture and in understanding DNA structure have enabled the development of transgenic species. Transgenic species have had a huge impact on ethical considerations in the area of genetics and reproductive technologies. For example, the production of Bt cotton raised many questions and opened up many possibilities in the area of agriculture, where the use of disease-resistant crops is viewed by producers as a huge advantage. *(8 marks)*

Question 19 (Total 4 marks)

(Answers will vary. The following table shows two types of answer.)

Type of cloning	*Process used*	*Example*
DNA cloning	Known genes are inserted into DNA of host	Used to produce unlimited amounts of identical copies of genes for study, production of useful protein, hormones, production of transgenic organisms (such as transferring the blue gene from a petunia into a carnation, creating the first light mauve carnation called Moondust) or insertion directly into plants such as tomatoes, cotton, corn of the Bacillus thuringiensis (Bt) gene which makes toxins that kill many pests
Whole organism cloning	A single cell is used to make an entire organism through recombinant DNA technology to create genetically identical offspring to the parent	Propagation of plants and production of animals for agriculture, such as fruit trees, sheep and pigs

(4 marks)

Question 20 (Total 8 marks)

Reproductive technologies that have the potential to influence evolution include artificial insemination, artificial pollination and whole organism cloning.

Each of these technologies has been developed to improve the reproductive potential of individuals that have been deemed more productive, usually for agriculture. This form of artificial selection has the potential to reduce biodiversity. A reduction in biodiversity makes species more vulnerable to change and therefore more at risk of declining population and extinction.

Chromosomes consist of DNA(the genetic code) combined with proteins such as histones. The key understanding about chromosome structure leading to these reproductive technologies is the fact that the chromosomes carry the genes and therefore the ability to transfer a nucleus (as is the case in somatic-cell nuclear-transfer cloning) from an organism with known characteristics into an enucleated egg cell, which means the subsequent development of the embryo and foetus should produce an individual genetically identical to the donor parent.

The behaviour of chromosomes in meiosis has assisted an understanding that the male gametes used in artificial insemination and pollination will still have some variation because of random segregation and crossing over. In addition, these gametes still require fertilisation with a female gamete, so there is an added source of variation. This means that cloning, if it becomes widespread, has greater potential to reduce biodiversity than artificial insemination and pollination. *(8 marks)*

OR

Our knowledge of chromosome structure has increased significantly over time, and has changed our understanding of chromosome structure and composition. Chromosomes are made of DNA. DNA is a double–helix-shaped nucleic acid. The strands of the helix consist of four different nucleotides, each made up of deoxyribose sugar, a phosphate molecule and a nitrogen base. Genes, which are identified with specific traits/characteristics, are coded within the DNA on the chromosomes. These genes have the potential to be isolated.

Reproductive technologies have moved forward from simple selective breeding for specific desirable characteristics to more complex cloning and transgenics. Through the physical manipulation of an organism's genome it is possible to change the individual gene sequence and to create an organism with desirable foreign genetic material. In these transgenic organisms, that material can be passed on from one generation to the next. Transgenic wheat has been developed to enhance crop yields, which can provide extra food in places where it is in short supply. Strawberries have been developed that are frost-resistant, so they can now be grown outside their normal growing season. This was made possible by inserting a gene from a salmon into the strawberries.

The cloning of agricultural species, such as fruit trees and livestock, reduces the natural variation within the gene pool of these organisms. Evolution relies on random variation and a struggle for existence, selecting the most favourable characteristics. New technologies such as cloning and transgenics mean that natural selection no longer drives evolution. They alter the path that nature would otherwise have taken. The technological advances that are being used to benefit society are possibly reducing the natural biodiversity that would have otherwise been created through evolution. *(8 marks)*

Question 21 (Total 4 marks)

Banana plants reproduce asexually, which means they have little genetic diversity because asexual reproduction does not result in the variation of offspring from the parents. Hence the genetic diversity of banana plants currently in existence will not be different to that of the originally introduced plants. Cloning is a process that produces genetically identical offspring to the parents. Thus if the banana plants were to be cloned, there would be no impact on the diversity of the banana plants; they would remain genetically identical to the parent plants. *(4 marks)*

Question 22 (Total 3 marks)

(a) Example: UV radiation *(1 mark)*

(b) UV radiation is greater in areas where the ozone is thinner, resulting in greater exposure and a higher incidence of skin cancer in those areas. *(2 marks)*

Question 23 (Total 8 marks)

(a) Bt cotton is an example of a transgenic species. It is used in the cotton industry as it increases the resistance of the cotton plant to a variety of pests, decreasing reliance on chemical pesticides. *(2 marks)*

(b) A transgenic species is produced by the transfer of a desired gene into the DNA sequence of another organism. One way of doing this is by using 'particle guns' to transfer DNA-coated microscopic pellets directly into the animal or plant, the desired gene having been isolated by the use of restriction enzymes from another organism and cloned by using a plasmid in bacteria. Once the gene is introduced into an individual, it becomes part of the organism and offspring are produced with the new gene being part of their DNA. *(2 marks)*

(c) Ethical issues arising from the use of transgenic species include:

- The selection of desired characteristics: who decides what is appropriate to introduce into another organism?
- What is the purpose of the introduction/modification?
- Who will benefit from the new species?
- The costs involved.
- The effects on the environment.
- The containment of the transgenic species in the environment.
- The effects on the food chain, including the long-term effects on human consumption.

(4 marks)

Question 24 (Total 3 marks)

'All mutations are harmful' is a very broad generalisation and is incorrect. A mutation describes any changes that may occur in an organism's DNA sequence. Changes can be harmful as well as beneficial to an organism. Without these changes occurring, the process of natural selection would not be able to take place, and organisms would not be able to survive in a changing environment. Hence the statement is not true. *(3 marks)*

Question 25 (Total 8 marks)

(a) Gene splicing can be carried out by isolating the required DNA, then cutting using an appropriate restriction enzyme. This can then be ultimately recombined by the use of a joining enzyme or ligase. *(2 marks)*

(b) If the fluorescent fish were accidentally released into the environment it might have significant impact on the environment due to the ability of the fish to mate with non-fluorescent fish, producing viable offspring. The resulting fish might have an adverse effect on the environment; for example, because they might be able to scare off predators, causing a significant increase in number. This increase in the number of fish without a natural predator would upset the food chain by placing stress on the food source, possibly depleting it. *(2 marks)*

(c) Different groups in society have varying opinions regarding the production of fluorescent zebra fish because the long-term effects on the environment are not known. While it is advantageous to be able to manipulate the genetic make-up of an organism to create desired characteristics, where these are not normally present, into the gene pool for the organism, the long-term implications must be considered.

On the surface, a fluorescent zebra fish may seem rather harmless for its intended purpose in a domestic fish tank, with short-term gains in terms of economic prosperity for the retailer/supplier being significant ('What a novel pet!'). It might not be so if it was released into the environment. If this was the case, the effects on the food chain might be significant and potentially cost far more in overcoming these effects.

Another significant concern for some groups in society relates to the issues dealing with the potential of the technology. Where will it stop? A fluorescent zebra fish is one thing, but what will be next? Who makes the decisions regarding the production of such organisms? *(4 marks)*

Question 26 (Total 3 marks)

(a) Cutting of DNA using restriction enzymes *(1 mark)*

(b) Ligation of enzymes involves the matching up of segments of DNA using the enzyme DNA ligase, which acts as an enzyme glue. In this process the enzyme DNA ligase is used to join the 'sticky' ends of the matching DNA segments. *(2 marks)*

Question 27 (Total 5 marks)

(a) One difference between RNA and DNA is the fact that RNA is single stranded and DNA is double stranded. *(1 mark)*

(b) Polymerase chain reaction (PCR) can be used to make copies of genes for analysis. *(1 mark)*

(c) The steps are: Step 1, denaturation, in which DNA is uncoiled; Step 2, annealing, in which primers attach to the DNA template; Step 3, extension, in which the gene is copied and the new copy extended. *(3 marks)*

Question 28 (Total 7 marks)

(*Note:* A number of products that are useful to humans have been produced via applications of modern biotechnology. Several will be described below. When a plural (e.g. 'applications', 'products') is used in a question, it is important to ensure that more than one example is in the answer.)

Modern biotechnology has produced a number of beneficial products in the areas of forensic science, medicine, animal and plant biotechnology and aquaculture. In medicine, examples include gene delivery via nasal spray, and the synthetic production of insulin. Plant and animal biotechnology has produced monoclonal antibodies and recombinant vaccines, while aquaculture has been responsible for the production of pharmaceuticals and marine farming.

Another useful biotechnology application that has been useful to humans involves the production of *monoclonal antibodies*. Antibodies are a component of the immune response, the third line of defence against infectious, disease-causing organisms.

Antibodies are specific, and confer an ongoing protection against the same antigens as a result of memory cells. As a result, antibodies have played a significant role in the effectiveness of vaccines. Antibodies can also be used to detect the presence of bacterial or viral products. However, the challenge to the effective use of antibodies is the production of large quantities of pure antibodies. Desired antibodies used to be collected from a laboratory host animal that had been injected with the particular antigen. Its blood would only yield small quantities of the desired antibodies as well as other undesirable substances.

A biotechnology solution to the problem is to identify natural cells that produce the desired antibodies and another type of cell that grows continuously (e.g. tumour cells). Biotechnology allows us to form a hybrid of these two cells, called a hybridoma. Effectively, we have a cell that grows and divides continually while retaining its antibody-producing function. It is an antibody production factory. These cells are called monoclonal because they originated from a single, fused cell. An example of monoclonal antibody technology is the fusing of bone marrow tumour cells (myeloma), which can be grown in pure cell culture, with mammalian spleen cells (producing a specific antibody). The antibodies produced by these monoclonal cells are much purer and more effective than traditional drugs in combating disease. The antibodies attack only the specific antigens, reducing the risk of side effects. *(7 marks)*

Question 29 (Total 6 marks)

(a) Each codon provides information for one amino acid, which is synthesised into polypeptides at the ribosomes. The genetic code determines the order of the amino acid in the polypeptide or protein, which acts as functional molecules such as hormones or enzymes. Genetic information is first synthesised from a DNA template (transcription) and the protein is then synthesised from a RNA template, called translation. If interference occurs during the translation process, where complementary mRNA is formed, this will result in a different polypeptide or protein, having a significant impact on the function of the cell. These changes can be deleterious to a cell, where the cell causes faulty function, or beneficial, resulting in differences that promote function or survival. This is central to natural selection.

RNA interference has an important role in defending cells against parasitic genes and in directing development, as well as gene expression in general. *(2 marks)*

(b) New biotechnology applications in the treatment of diabetic patients have been developed, including the production of insulin through gene technology, a process that utilises the human insulin gene and inserts it into the patient's own cells. The gene is inserted into liver cells rather than damaged pancreatic tissue, as the liver cells are similar to the beta cells that produce insulin. The new gene or section of DNA (human rather than from another animal) is inserted into the patient's liver cell nuclei. The translation of this section of DNA into complementary bases then involves reverse transcription, which will create a copy of RNA that can then produce the new polypeptide or protein. This, in turn, will complete the new strand and form the insulin. *(4 marks)*

Question 30 (Total 4 marks)

Step 1: Gene is 'cut out' of one species using restriction enzyme.

Step 2: Circular piece of DNA (plasmid) is removed from a bacterial cell and cut open with restriction enzyme (e.g. from *E. coli*).

Step 3: 'Cut out gene' is mixed with bacteria plasmid in a test tube, where the ends match up (recombining with the gene spliced into it) and are stuck together using DNA ligase.

Step 4: 'Transformation into bacteria'. Plasmid containing recombinant DNA is re-inserted into the bacteria by mixing it with calcium chloride or the use of electroporation.

Step 5: Introduced gene will now be expressed by the bacterial cell.

(4 marks)

Question 31 (Total 5 marks)

(a) A *qualitative* observation is one that describes the features of something, such as 'ants remove any other plants that may spring up among their crop'. A *quantitative* observation is one that uses a measurement; for example, 'a ten-year study has shown that the average size of the grass seeds increased by 5%'. *(2 marks)*

(b) The ant activity could be described as biotechnology because the ants 'retain some seeds for next year's crop'. This implies they are actively selecting some seeds (the bigger ones) for planting the following year and, as a result, over a ten-year period, the average size has increased. The ants have manipulated the seeds to their benefit. *(3 marks)*

Question 32 (Total 7 marks)

These experiments are very significant for medical technology and assisting patients with diseases that can be linked to specific genes. In particular, the XIST gene may provide a treatment for those that suffer from trisomy/multiple X diseases associated with the X chromosome. If the inserted XIST gene silences the X chromosome, the symptoms that manifest themselves with the extra X chromosome may be diminished.

Such treatment is called gene therapy, which has been successfully carried out on patients with cystic fibrosis. In these patients the genes necessary for healthy lung function are inserted into the lung cells, using adenoviruses as vectors, delivered as inhalants. There has been varied success with this treatment due to problems associated with the size of the gene, lung cells being replaced relatively frequently and sometimes the carrier virus causes inflammation of the lung tissue. However, there has been success with cancer suppressor cells administered via adenovirus. In conjunction with anti-cancer drugs these treatments have resulted in the regression of tumours.

The use of retroviruses for gene therapy has had limited success but research in this area continues due to the possible benefits if the technology improves, which is only a matter of time. However, such treatment may pose ethical considerations. If carried out as somatic gene therapy, the functional gene would be inserted into the appropriate body cells, and as such would not be passed on. If the treatment involved germ line therapy (manipulation of gene in a gamete), the change could be passed on. This gives rise to ethical concerns, with questions arising regarding the nature of treatment and the possible implications for the individual in the future and the impact on the wider community. *(7 marks)*

Question 33 (Total 8 marks)

(This sample answer may have more information, such as examples, than would be required in an HSC Examination.)

	Point mutations	*Chromosomal mutations*
Causes	Change of one nucleotide base in DNA occurring during replication before mitosis or meiosis; may result from exposure to mutagens (electromagnetic, chemical or natural)	Problems in chromosome behaviour during meiosis, mitosis or the action of mutagens such as ionising electromagnetic radiation (gamma rays, x-rays and some ultra violet rays)
Processes	Substitution, deletion or insertion of a nucleotide; deletions and insertions cause a frameshift as they alter the sequence of codons in mRNA that transcribes the mutated DNA.	• Duplication—a segment of chromosome is duplicated and incorporated into the chromosome. • Deletion—a segment of chromosome is removed. • Inversion—a segment of chromosome breaks off, changes orientation by 180° and is reinserted into the chromosome. • Translocation—a segment of chromosome breaks off and joins a chromosome on another homologous pair. • Aneuploidy (an entire extra or missing chromosome) results from non-disjunction of homologous chromosomes during cell division, e.g. trisomy. • Polyploidy—more than two haploid sets of chromosomes, e.g. triploidy

Question 33 continues

Question 33 (continued)

	Point mutations	*Chromosomal mutations*
Effects Both types of mutations are a source of genetic variation and in the long term have allowed adaptation and evolution of species, even though in some cases they may be harmful. Germ-line mutations are passed on to offspring, and somatic mutations may cause diseases such as cancer, but only effect the individual. Point mutations can occur in 'coding' or 'non-coding' parts of the DNA. Mutations in 'non-coding' DNA may influence gene regulation.	Substitution mutations in 'coding' DNA may result in no alteration of the amino acid sequence in the resultant polypeptide (i.e. a silent mutation), or the change of one amino acid that results in an alteration of function of the polypeptide (missense mutation). If the substituted change results in the production of a stop codon, the result will probably be a failure in the production of the polypeptide (i.e. non-sense mutation). Insertions and deletions that cause frameshifts in the mRNA generally result in non-sense mutations that prevent the formation of functional polypeptides.	When they occur in germ-line cells they may be fatal or result in severe disorders of offspring, e.g. Duchenne muscular dystrophy, Down syndrome or XX male infertility. Triploidy (and other forms of polyploidy with uneven multiples of haploid sets) results in sterility as meiosis cannot occur. Polyploidy is common in plants, some fish, frogs, salamanders and leeches.

(8 marks: One mark each for causes and processes and two marks each for effects)

Question 34 (Total 6 marks)

Many chemicals that are mutagens have been developed to inhibit the cell division of cancerous cells (i.e. they are chemotherapy drugs). One group of chemical mutagens belong to the family of mustard gases used as a chemical weapon during World Wars I and II. These compounds add small hydrocarbon molecules to all parts of the DNA molecule, causing a range of mutations in the DNA. Cigarette smoke contains a range of chemical mutagens such as polycyclic aromatic hydrocarbons (PAH) that cause mutations by attaching to various parts of the DNA.

Naturally occurring mutagens include some viruses called oncoviruses. These viruses take over the host's genome and bring about changes in the DNA. The human papilloma virus that causes cervical cancer is an oncovirus. Other naturally occurring mutagens are chemicals such as aflatoxin and flavonoids that are produced by moulds and in some plants. Reactive oxygen species (ROS) are produced by mitochondria but are produced in larger amounts as a result of exposure to pollutants, radiation and in some cases environmental stress. ROS form cross links in DNA, attach to bases and cause breaks in DNA strands. *(6 marks)*

Question 35 (Total 5 marks)

Whole organism cloning has different methods and outcomes in plants to animals. Many plant propagation techniques such as growing rhizomes, bulbs, corms and cuttings have been used widely over a long period to result in plant clones. Obtaining reliable information from both print and online sources was easily confirmed by the consistency of information. Animal cloning is more recent, controversial and requires high levels of skills and technology, so selecting well-reputed sources such as the Roslin Institute (University of Edinburgh) was important. It was also important to ensure that information was current because changes have occurred in both techniques and results over the last 20 years.

The effectiveness of plant cloning is easily determined by the high level of its use and the benefits to agriculture and horticulture that have ensued. Early reports on animal cloning suggested major problems with premature aging of the clones and susceptibility to disease. Public acceptance of animal food sources that have been cloned is also a factor that has limited effectiveness. Care had to be taken to ensure that information from both sides of some arguments about the use of animal clones was considered. *(5 marks)*

CHAPTER 3

Module 7
Infectious Disease

Part A Objective-response questions

1 The following diagram summarises the steps of an experiment similar to that carried out by Louis Pasteur, which identified microbes as agents of decay.

Step 1 Two swan-neck flasks are filled partially with equal volumes of beef broth.

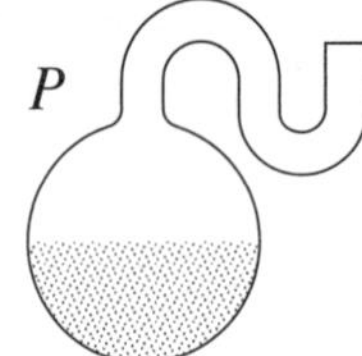

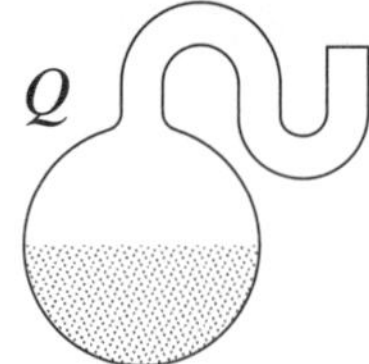

Step 2 The broth is boiled for at least 20 minutes.

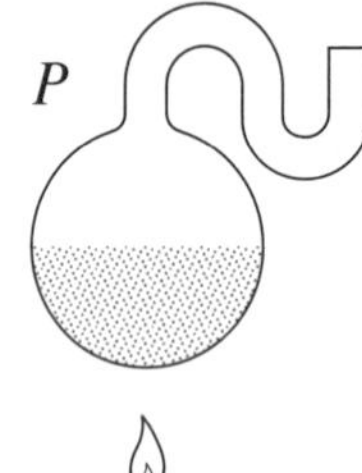

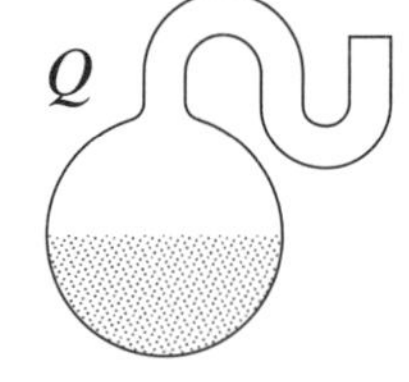

Step 3 The neck of Flask *P* is left intact whereas the other is broken.

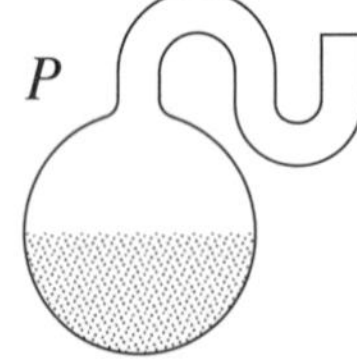

Step 4 The flasks are observed two weeks later for evidence of decay

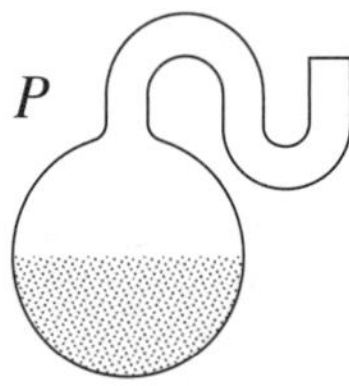

No decay

Decay present

Which of the following statements best explains the results obtained?

(A) There were no microbes in the air around Flask *P* at Steps 3 or 4.

(B) There were no microbes in Flask *P* at the beginning of the experiment.

(C) Microbes in Flask *Q* were not all killed by boiling, and multiplied following the cooling down of the flask.

(D) Any microbes present in both Flasks *P* and *Q* were killed by the boiling process, and only Flask *Q* allowed microbes to re-enter.

(q12, 2003 HSC)

2 What is the function of T-helper cells?

(A) Initiation of inflammation

(B) Phagocytosis of bacteria and viruses

(C) Promotion of B cell and T cell activity

(D) Production of specific antibodies against pathogens

(q14, 2003 HSC)

3 Eight sick animals were found to be suffering from the same symptoms. Blood tests showed that they were infected with the same type of bacterium.

Which of the following strategies would be the best to determine if this particular type of bacterium is the cause of the disease?

(A) Find other animals with the same symptoms. Attempt to isolate the same type of bacterium from their blood.

(B) Inject blood from animals with the symptoms into suitable host individuals. If they develop the same symptoms, this proves that this type of bacterium caused the disease.

(C) Use bacteria cultured from the blood of the animals with these symptoms to infect suitable host individuals. If they develop the disease, attempt to isolate the same type of bacterium from their blood.

(D) Treat all eight animals with an antibiotic known to kill this type of bacterium. They will recover if this type of bacterium is the cause of the disease.

(q5, 2001 HSC)

4 The following paragraph describes a body response.

> The response is protective, and it makes nearby blood vessels leak. Plasma and white cells move into the affected area, diluting and destroying the infectious agent.
>
> This is why the infection site swells, reddens and feels hot. Although we tend to think of this response in terms of annoyance, soreness and pain, it is actually a beneficial response.

What response does this paragraph describe?

(A) Inflammation

(B) Cell differentiation

(C) The action of antibodies

(D) The activation of helper T cells

(q6, 2002 HSC)

5 Which of the following pairs prevent entry of pathogens into the human body?

(A) The skin and phagocytosis

(B) The skin and chemical barriers

(C) Inflammation response and phagocytosis

(D) Inflammation response and chemical barriers

(q3, 2009 HSC)

6 The potential for disease to spread through animal populations in intensive farming is heightened because the animals are kept close together.

A disease has been identified in animals in one enclosure on a farm.

Which procedure would best prevent the spread of the disease to animals in other enclosures on the farm?

(A) Isolate diseased animals from healthy animals then vaccinate all healthy animals.

(B) Vaccinate all animals so that healthy animals do not develop the disease and spread it further.

(C) Move the diseased animals into another enclosure to quarantine them from the healthy animals.

(D) Wash all animals with antiseptic solution so that the pathogen causing the disease cannot be spread from diseased animals to healthy animals.

(q4, 2009 HSC)

7 What feature of prions distinguishes them from all other types of pathogens?

(A) Prions are not cells.

(B) Prions do not contain DNA.

(C) Prions do not contain nucleic acids.

(D) Prions cannot reproduce outside a host cell.

(q17, 2010 HSC)

8 A model of a virus is shown.

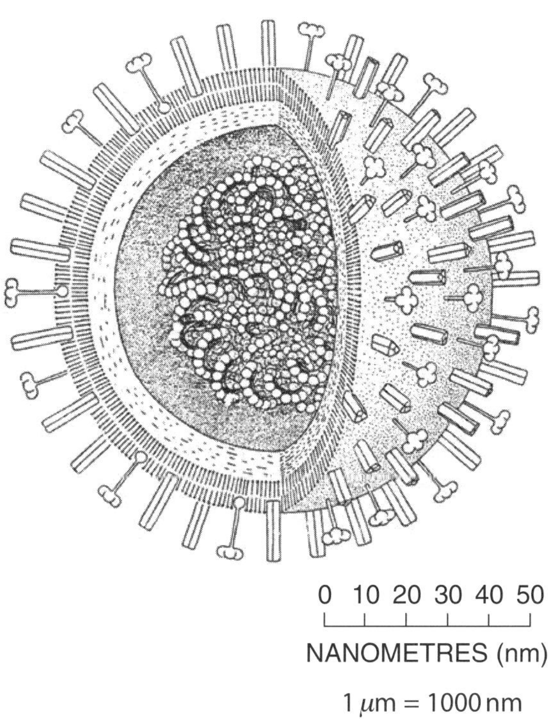

What is the approximate diameter of this virus?

(A) 13 cm

(B) 13 nm

(C) 130 μm

(D) 0.130 μm

(q18, 2010 HSC)

9 The diagram illustrates an immune response.

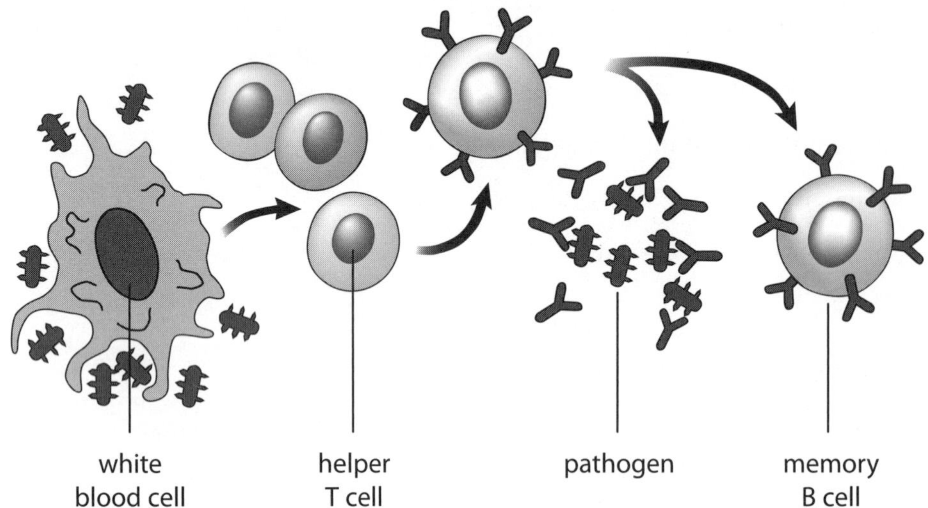

Below is a list of statements, each describing a step in the immune response.

1 Antibodies are produced to immobilise the pathogens.

2 B cell is activated by a helper T cell.

3 Helper T cells are activated by the white blood cell.

4 Memory B cell is ready to respond to further infections.

What is the correct sequence of events?

(A) 2, 3, 1, 4

(B) 2, 3, 4, 1

(C) 3, 2, 1, 4

(D) 3, 4, 2, 1

(q20, 2010 HSC)

10 Which of the following shows the correct sequence of steps in the interaction between T cells (lymphocytes) and B cells (lymphocytes)?

	Step 1	*Step 2*	*Step 3*
(A)	T cells interact with antigens presented by macrophages	Activated T cells differentiate into helper T cells	Helper T cells produce chemicals that cause B cells to multiply
(B)	T cells interact with antigens presented by macrophages	Activated T cells differentiate into B cells	B cells produce chemicals that cause T suppressor cells to multiply
(C)	B cells interact with antigens presented by macrophages	Activated B cells differentiate into plasma cells	Plasma cells produce chemicals that cause T cells to multiply
(D)	B cells interact with antigens presented by macrophages	Activated B cells differentiate into B memory cells	B memory cells produce chemicals that cause macrophages to multiply

(q16, 2011 HSC)

11 Which list shows pathogens in order of increasing size?

(A) Bacterium, prion, virus, protozoan, macroparasite

(B) Macroparasite, prion, protozoan, bacterium, virus

(C) Prion, virus, bacterium, protozoan, macroparasite

(D) Virus, prion, bacterium, macroparasite, protozoan

(q9, 2011 HSC)

12 The diagram shows a pathogen called *Giardia*. What type of pathogen is *Giardia*?

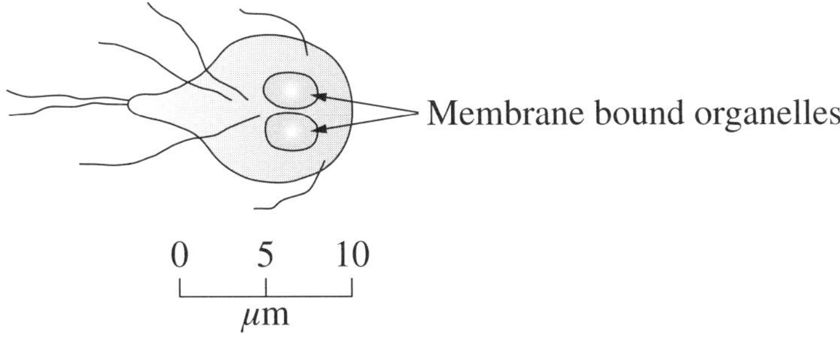

(A) Bacterium

(B) Prion

(C) Protozoan

(D) Virus

(q4, 2012 HSC)

13 How do vaccinations prevent disease?

(A) They increase the inflammation process.

(B) They enable the infected cells to seal off the pathogen.

(C) They increase the number of antibodies against the pathogen.

(D) They decrease the number of antigens that trigger the immune response.

(q6, 2012 HSC)

14 Why is it important to continue research into new antibiotics?

(A) New prion diseases have been recently discovered.

(B) Resistant bacteria have evolved from the overuse of antibiotics.

(C) Viral infections require a broad range of antibiotics for eradication.

(D) New diseases are discovered regularly and all require new antibiotics.

(q7, 2012 HSC)

15 Antibodies are proteins that:

(A) break down pathogens.

(B) bind with a specific antigen.

(C) catalyse biochemical reactions.

(D) are produced by T cells to kill disease-causing viruses.

(q2, 2013 HSC)

Part B Short-answer questions

Marks

Question 16 (4 marks)

Further avian influenza outbreaks in Indonesia

Avian influenza has been found closer to Australian shores, with more outbreaks in Indonesia.

The spread of the disease has slowed across most Asian countries, but in Indonesia it has spread as far south as Lombok and West Timor.

A representative for Australian chicken growers says the new outbreaks don't mean Australia's chances of contracting the disease are any greater, but high quarantine measures remain in place.

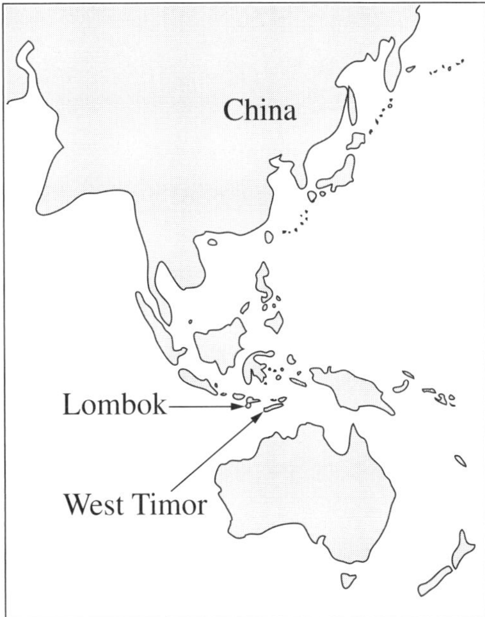

Using avian influenza as an example, evaluate the effectiveness of quarantine measures in Australia. **4**

(11 lines) (q19, 2004 HSC)

Question 17 (3 marks)

(a) Identify ONE type of *T* lymphocyte. **1**

(1 line)

(b) Distinguish between the functions of *B* cells and *T* cells. **2**

(4 lines) (q20, 2004 HSC)

Question 18 (4 marks) **Marks**

To study the effect of an antibiotic on three strains of bacteria (*A*, *B*, *C*), agar plates were set up as shown.

	Agar plate 1	*Agar plate 2*	*Agar plate 3*
Method	Surface covered with bacterium *A*	Surface covered with bacterium *B*	Surface covered with bacterium *C*
	Incubation 37°C, 48 hours		
Result			

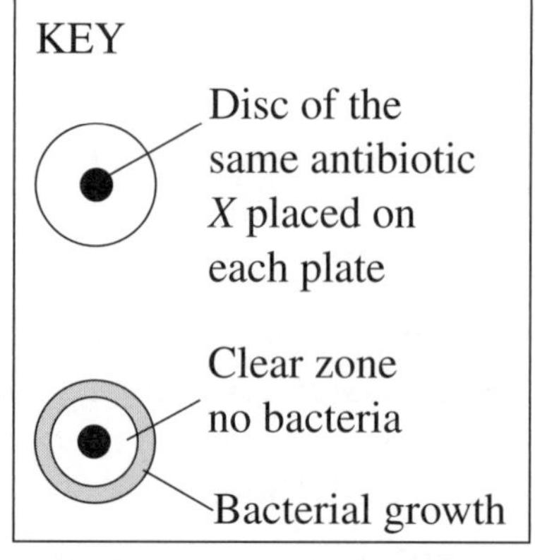

The plates were incubated at 37 °C for 48 hours. The diagrams of plates 1, 2 and 3 depict the results.

(a) Write a conclusion to the experiment. **1**

(2 lines)

(b) Identify ONE safe work practice used to minimise risks associated with handling or identifying microbes. **1**

(2 lines)

(c) In humans, bacterial infections are often treated with antibiotics. Explain why the complete course of antibiotics should be taken, even if the symptoms of infections have disappeared. **2**

(4 lines) (q27, 2004 HSC)

1

Question 19 (8 marks) **Marks**

Consider the following information.

> An Australian pathologist from Perth, Robin Warren, discovered that the bacterium *Helicobacter pylori* was associated with inflammation and ulceration of the stomach and duodenum. Barry Marshall, a young clinician, was interested and joined Warren's research.
>
> Previous treatments of these conditions were based on the assumption they were caused by stress and lifestyle. Treatment was based around the stopping gastric acid secretion. Relapses were frequent until these researchers showed that killing the bacteria resulted in long-term recovery.
>
> Warren and Marshall's investigations included:
>
> - using a microscope to look at prepared slides of ulcerated stomach tissues;
> - using a flexible endoscope to look into the stomach of patients with stomach ulcers and gastritis (localised or general inflammation of the stomach);
> - using staining techniques to determine the possible presence of bacteria in stomach tissue.
>
> To demonstrate, Warren checked that Marshall's stomach contained no *Helicobacter pylori*. Marshall then swallowed a dose of the bacteria, triggering symptoms of gastritis.
>
> (Source: Summarised from Pincock, S. (2005). Nobel Prize winners Robin Warren and Barry Marshall. *The Lancet* 366(9495): 1429. DOI 10.1016/S0140-6736(05)67587-3)

Assess the procedures that Warren and Marshall used to identify and confirm their conclusion about the pathogen that caused stomach ulcers and gastritis. **8**

(26 lines) (Updated q26, 2006 HSC)

Question 20 (5 marks)

Fungicides are chemicals that are used to treat fungal diseases such as rust in plants.

Rust symptoms include orange-brown patches on the underside of leaves.

A new brand of fungicide claims to successfully treat rust disease in plants.

You designed an experiment to test this claim. Other students also performed your experiment and gave you their results.

(a) Write a valid procedure for your experiment. **4**

(12 lines)

(b) How would you assess the reliability of your experimental design? **1**

(2 lines) (q23, 2014 HSC)

Question 21 (3 marks) **Marks**

(a) The diagram shows a process that is a part of the immune response.

What is the name of the process? 1

(1 line)

(b) Outline how inflammation contributes to the immune response. 2

(4 lines) (q21, 2014 HSC)

Question 22 (8 marks)

Evaluate the contributions made by both Louis Pasteur and Robert Koch to our present understanding of the causes and possible prevention of infectious diseases. 8

(25 lines) (q27, 2002 HSC)

Question 23 (3 marks)

Public health programs, pesticides and genetic engineering are used to control and/or prevent disease.

Using an example, explain how ONE of these strategies has been used to control or prevent disease within the community. 3

(6 lines) (q22, 2004 HSC)

Question 24 (2 marks)

'Black spot' is an infectious apple disease that occurs in New Zealand and makes apples unsuitable for sale.

Describe a method to prevent the spread of the disease into Australia. 2

(4 lines) (q24, 2006 HSC)

Question 25 (3 marks) **Marks**

A student working in a restaurant kitchen is required to wear disposable gloves and hat when preparing food.

(a) Explain how this practice assists in the control of disease. **2**

(4 lines)

(b) Identify another hygiene practice that reduces the risk of infection. **1**

(1 line) (q16, 2007 HSC)

Question 26 (6 marks)

(a) A gardener noticed a red and swollen area on his arm that had received a deep scratch from a thorn on a plant. **1**

Identify the most likely human defence adaptation that caused these symptoms.

(1 line)

(b) After a number of days, the gardener's arm remained red and swollen, so he visited his doctor, who prescribed an antibiotic to treat the infection. **1**

Why did the doctor prescribe an antibiotic to treat the infection?

(2 lines)

(c) The gardener should have taken the antibiotic for ten days but stopped after five days because the arm was no longer red or swollen. **4**

Explain how this action by the gardener might lead to antibiotic resistance.

(8 lines) (q28, 2006 HSC)

Question 27 (6 marks)

(a) Name ONE example of a disease caused by a macro-parasite. **1**

(1 line)

(b) List TWO features of prions that distinguish them from protozoans. **2**

(2 lines)

(c) Most pathogens must first be transmitted to and enter the human body before they trigger an immune response. **3**

Relate this statement to a named infectious disease that you have studied.

(5 lines) (q19, 2007 HSC)

Question 28 (4 marks) **Marks**

A new product has been developed to kill pathogens in drinking water. **4**

Design an experiment to test the effectiveness of the product.

(10 lines) (q19, 2008 HSC)

Question 29 (3 marks)

The Tasmanian Devil is in danger of becoming extinct due to an infectious disease that causes facial tumours. The animals slowly starve and usually die within six months of showing tumours. Populations in the western third of Tasmania currently remain free of this disease.

Justify the steps you would take to ensure that a population of Tasmanian Devils remains disease free. **3**

(6 lines) (q26, 2008 HSC)

Question 30 (3 marks)

Construct a table to identify THREE types of T lymphocytes, and outline the role of each type in the immune response. **3**

(12 line spaces) (q22, 2009 HSC)

Question 31 (3 marks)

Complete the following table. **3**

Pathogen	*Distinguishing characteristic of the pathogen*	*Disease caused by this type of pathogen*
Bacteria		
Fungi		
Protozoans		

(q16, 2009 HSC)

Marks

Question 32 (6 marks)

Compare direct, indirect and vector transmission of infectious disease. **6**

(12 lines) (Sample question)

Question 33 (4 marks)

Explain the differences between innate and acquired immunity. **4**

(10 lines) (Sample question)

Question 34 (8 marks)

Compare the effectiveness of antivirals with antibiotics. **8**

(24 lines) (Sample question)

Question 35 (5 marks)

Describe the investigation carried out into the response of a named Australian plant to a named pathogen. **5**

(12 lines) (Sample question)

Module 7
Infectious Disease
Answers

Part A Objective-response questions

1 D This answer describes Pasteur's experiment correctly.

2 C Correct definition.

3 C This is the only option that correctly follows Koch's postulates in identifying the disease.

4 A The paragraph describes the second-line defence mechanisms.

5 B The skin and chemical barriers are associated with only the 'first line of defence' and hence with the prevention of entry of pathogens into the body.

6 A This option removes the healthy animals and minimises contact with the diseased animals. Vaccination will more effectively decrease the chance of the disease spreading than any other option.

7 C Nucleic acids make up DNA, so this is the better option.

8 D The diameter of the virus measures 130 nm or 0.130 μm.

9 C This option gives the correct sequence for the action of antibody mediated response.

10 A It is the helper T cells that cause the multiplication of B cells, resulting in the interaction between the cell types.

11 C Prions are approximately 10 nm, viruses 20–100 nm and bacteria 1–5 μm.

12 C It is a protozoan as it is a single-celled organism with membrane-bound organelles.

13 C Vaccinations, through the introduction of harmless/dead strains of the disease, enable the body to build up antibodies so that it can fight the disease should it be exposed to it.

14 B Due to the natural selection of bacteria, some antibiotics are ineffective and we have a problem with combating some severe bacterial infections.

15 B An antibody identifies antigens, which are specific markers on pathogens or foreign targets.

Part B Short-answer questions

Question 16 (Total 4 marks)

Quarantine measures have been used very effectively in Australia to prevent the entry of many pathogens, pests and disease-carrying organisms. Avian influenza is not a major risk to Australian farmers as the quarantine measures currently in place should be adequate to prevent the spread of the disease into Australia from our Asian neighbours.

Measures such as the restriction of entry to certain goods and products, inspection points at all ports, the destruction of affected goods, and quarantine isolation and fumigation have all so far effectively assisted in keeping disease (including foot and mouth, rabies and various plant diseases) out of Australia. However, they may not protect against the spread of disease by wild birds that may migrate naturally. *(4 marks)*

Question 17 (Total 3 marks)

(a) T lymphocytes: killer T cells, suppressor T cells, helper T cells, cytotoxic T cells. *(1 mark)*

(b) The human immune system relies on both B cells and T cells. The T cells are used for B cell activation and phagocytosis (that is, a cell-mediated response), while B cells control the humoral response and produce antibodies to antigens present in the blood. *(2 marks)*

Question 18 (Total 4 marks)

(a) The antibiotic is most effective against bacteria *A* and not at all effective against bacteria *C*. *(1 mark)*

(b) One safe work practice used to minimise risks associated with handling or identifying microbes would include such procedures as wearing gloves.

Other examples are keeping plates sealed, washing down surfaces/benches with alcohol, the correct disposal of medium and any contaminated plates/inoculation equipment, etc., with high temperatures or appropriate chemical treatments, and washing of hands with recommended sterilisation wash. *(1 mark)*

(c) The completion of antibiotic courses is recommended, even once symptoms have disappeared, to ensure that all the bacteria are completely destroyed. This procedure minimises the chances of resistance build-up in the bacteria strains, thereby enabling the continued effectiveness of the antibiotic for the treatment of the bacterial infection in the future. *(2 marks)*

Question 19 (Total 8 marks)

(Answers will vary but should include a discussion of the method used by Warren and Marshall in terms of validity and reliability; that is, the use of numerous clinical cases for observation of symptoms, and testing for the presence of the bacteria over an extended period of time, enabling the experiments to be validated. An example is provided.)

The procedure was valid because they not only examined the stomachs of many patients but also sampled the stomach lining and examined the tissue under a microscope, detecting the presence of a similar micro-organism in each case, which was later confirmed.

The procedure they carried out essentially followed the rules for identification of disease set out by Robert Koch in the late 1800s.

Koch's postulates:

- The specific micro-organisms must be present in each case of the disease.
- The specific micro-organisms must be isolated from the host and grown in a pure culture.
- A potential host, when inoculated with the micro-organism, must develop the same symptoms as the original host.
- The specific micro-organism must be able to be isolated from the second host and identified as the same species as originally cultured.

First they isolated what they thought was the causative pathogen, present in every case observed: the bacterium *Heliobacter pylori*. Then they introduced the pathogen into an otherwise healthy individual: Marshall himself. Marshall then exhibited the same symptoms as the other patients examined, showing that it was indeed a bacterium that caused the gastric ulceration in the patients, not the previously thought causes of stress and poor hygiene habits.

In order to improve reliability of the experiment, the later step should have been repeated,

to ensure that introducing the *Heliobacter pylori* into a larger sample of healthy organisms would have consistently reproduced the same result. *(8 marks)*

Question 20 (Total 5 marks)

(a) Validity is achieved when only the independent variable is changed (in this case, the fungicide used). All other variables must be kept constant. The following steps outline a valid experiment to test the effectiveness of the fungicide.

1. Obtain a large sample (e.g. 60) of infected plants and an identical control group of plants that are not infected.
2. Plant in identical trays containing the same volume and type of soil. Water with the same volume of water at regular intervals of two to three days, using a measuring cylinder. (The trays, soil and water are controlled variables.)
3. Leave the plants in the same position, one that is appropriate to promote growth for that type of plant.

4 Spray half of each group of plants with the fungicide from a spray bottle, ensuring the plants are sprayed identically. Monitor the plants daily to see if the fungicide makes any difference to the growth of the infected and control groups.

5 Repeat these steps multiple times with the same fungicide to ensure reliability of results. *(4 marks)*

(b) The reliability of the experiment could be assessed by looking at how many times the experiment was repeated by different individuals, the sample size and the average of the results. *(1 mark)*

Question 21 (Total 3 marks)

(a) Phagocytosis *(1 mark)*

(b) Inflammation is a non-specific immune response. The body brings blood and fluid to the location of damaged cells or infection, causing it to redden, swell and warm.

The extra blood carries phagocytes to destroy the pathogen, enables clotting to occur and prevents pathogens spreading. *(2 marks)*

Question 22 (Total 8 marks)

Both Louis Pasteur and Robert Koch contributed significantly to the understanding of the cause and prevention of infectious disease in modern medicine. Each contributed independently by way of research in different areas.

Pasteur identified that micro-organisms develop from pre-existing micro-organisms rather than from spontaneous generation. He developed sterilisation as a method of destroying pathogenic micro-organisms. This enabled further links between micro-organisms and disease to be made, including the destruction of the micro-organisms and the understanding for the need for basic hygiene, sanitation and water treatment.

Koch is attributed with the identification of a series of postulates that could be used to identify a micro-organism and determine whether it was responsible for a particular disease.

This contribution led to the identification of a range of specific disease-causing micro-organisms and subsequently to their treatment.

Together the work of these men is directly responsible for the procedures used in identifying and treating diseases today, whether by eliminating the micro-organisms from food and water or isolating the causative agent of a disease, making it possible to develop a treatment or vaccinations to prevent its spread. *(8 marks)*

Question 23 (Total 3 marks)

Public health programs are used to educate wider general populations regarding the dangers of particular diseases or practices, enabling people to identify symptoms and take preventative action to reduce the incidence. Programs use advertisements and pamphlets.

One such program includes the 'Slip, slop, slap' campaign, used to raise awareness of the danger of exposure to harmful UV radiation from the Sun. By educating the population about the dangers of going outside on warm days without such protection as clothing, hat or sunscreen, the incidence of cancers such as melanoma can be reduced in future generations. Together with regular screening and early intervention, the incidence of skin cancer may be reduced.

Examples of other programs: AIDS and sexually transmitted diseases campaign, early childhood immunisation, 'Life. Be in it.' campaign. *(3 marks)*

Question 24 (Total 2 marks)

Australian quarantine measures prevent the spread of the black spot disease into the country by regulating the quality of fruit/apples and inspecting produce prior to entry.

Suppliers are required to conform to Australian standards, including preventing exposure to certain chemicals and checking produce for a variety of diseases before it leaves the point of origin. Australian inspectors carry out checks/tests on arrival, thereby minimising the possibility of such diseases entering the country by banning imports of affected fruit/apples. *(2 marks)*

Question 25 (Total 3 marks)

(a) A student working in a restaurant kitchen is required to wear disposable gloves and hat when preparing food so that disease transmission is reduced. Hands can transmit many diseases if they have not been washed thoroughly. Hair can transmit skin diseases as well as various parasites that can fall onto the food. *(2 marks)*

(b) Other hygiene practices that reduce the risk of infection can be any of the following:

- Handwashing
- Refrigerating perishable foods rather than leaving them at room temperature
- Using clean utensils and plates
- Wearing a face mask to prevent fine droplets produced by sneezing/ coughing from transmitting disease
- Covering any open wounds. *(1 mark)*

Question 26 (Total 6 marks)

(a) The most likely human defence adaptation that caused the symptoms was the second line of defence or inflammation response. *(1 mark)*

(b) The doctor prescribed an antibiotic for the gardener because the swelling was caused by the body's reaction to a bacterial infection. *(1 mark)*

(c) Not taking the antibiotic beyond five days can result in the bacteria developing a resistance to the antibiotic because it is possible that not all the bacteria will have been destroyed by the antibiotic at this time. Darwin's theory of natural selection states that within any population, variation exists, and those organisms best suited to their surroundings will survive to reproduce and pass those characteristics on to offspring. It is possible that some bacteria are genetically different from those destroyed and they will continue to reproduce, passing on this variation and thereby creating a species that is not affected by the antibiotic in the future. *(4 marks)*

Question 27 (Total 6 marks)

(a) Diseases caused by a macro-parasite include: fascioliasis (liver fluke disease), bilharzia (parasitic worms spread by snails), schistosomiasis (blood fluke disease), taeniasis (tapeworm disease), lice infestation (lice), and scabies (itch mites). *(1 mark)*

(b) Prions are proteins that have been altered from the normal shape while remaining chemically the same and they cannot exist independently outside of a living organism, while protozoans are free-living and single-celled eukaryotic organisms.

Prions are extremely resistant to heat and chemical agents, unlike protozoans. *(2 marks)*

(c) Many infectious diseases may be discussed, for example, malaria. Malaria is transmitted by the *Anopheles* mosquito. The causative agent is any one of four different microscopic sporozoan protozoans from the genus *Plasmodium* that can cause malaria via the mosquito vector. The infective *Plasmodium* sp. must first enter the body and get past the first line of defence. This is done by a mosquito vector that injects an anticoagulant containing saliva into the animal host through the skin while sucking the host's blood. Transmission occurs when the saliva containing the *Plasmodium* sp. is injected into the bloodstream. It then invades the liver cells, where asexual reproduction occurs. Consequently, red blood cells are affected, causing them to burst, releasing toxins into the body of the host. Once the parasite enters the red blood cells, antibodies are produced, but surface antigens on the cell membrane of the *Plasmodium* sp. are varied and hence the host immune response is useless as the antibodies do not recognise the antigens. *(3 marks)*

Question 28 (Total 4 marks)

(Answers will vary, but should include the key elements of experimental design. Examples are provided.)

- Identification of dependent/independent variables, keeping the independent variables (e.g. temperature, water source, volume) constant so that any change is attributable to the effectiveness of the product
- Identification of a method of testing for the pathogen
- Large sample size/repetition of experiment
- Logical sequence of events
- Identification of risks. *(4 marks)*

Question 29 (Total 3 marks)

(Answers will vary, but must convey the idea of quarantine measures being applied to restrict the transmission of the disease throughout Tasmania. An example is presented.)

To keep the western third of Tasmania free of facial tumour disease, the area must be quarantined such that the movement of Tasmanian Devils into and out of the area is restricted.

While it may be difficult to completely prevent movement into and out of the area, because fencing the area to prevent the movement of the animals themselves may be not be practical, checkpoints need to be established on routes leading into the area, so that contaminated Tasmanian devils—or animals in contact with them, such as domestic animals—are not transported into the area. Tagging Tasmanian Devils into and out of the area may provide valuable information regarding their normal movement. Establishing isolated, disease-free 'insurance' populations of Tasmanian Devils is another strategy. *(3 marks)*

Question 30 (Total 3 marks)

(Answers could include any of the following cell types.)

Roles of three types of T-lymphocytes		
Cytotoxic T cells	*Memory T cells*	*Helper T cells*
Destroy any cells that carry foreign antigens, removing foreign protein such as bacteria or transplant tissue	Recognise any antigen when it reappears, enabling antibody production for the particular antigen	Secrete interleukens that regulate the function of both *cytotoxic T cells* and *B cell* function

Other T lymphocytes could include: inducer T cells, natural killer cells and suppressor T cells. *(3 marks)*

Question 31 (Total 3 marks)

(Sample answers could include the following.)

Pathogen	*Distinguishing characteristic of the pathogen*	*Disease caused by this type of pathogen*
Bacteria	Moneran, single-celled, procaryotic	Meningiccocal, tetanus, pneumonia, tuberculosis
Fungi	Parasitic, microscopic tubular filaments/threads, hyphae, spores	Ringworm, athletes' foot, rusts, moulds, blight
Protozoans	Single-celled, eucaryotic	Malaria, giardia, dysentry

(3 marks)

Question 32 (Total 6 marks)

The transmission of an infectious disease results in diseases spreading within families or communities or even becoming epidemics and pandemics. Direct transmission of pathogens results from physical contact between hosts or infected droplets carried up to a metre through the air (droplet transmission). Saliva, mucus, blood and other body secretions and fluids can be the source of direct transmission. Direct transmission is often called person-to-person transmission. Direct transmission can occur between mother and offspring via the placenta, during childbirth or in breast milk.

Indirect transmission occurs when surfaces (fomites) such as doorknobs or medical equipment retain contaminated material and transfer it to other hosts. Indirect transmission is also the result of airborne transmission of infected droplets from coughing or sneezing that are carried over a metre in distance.

Vector transmission results from living organisms transferring the pathogen, either by being infected themselves or by mechanically transferring the pathogen without becoming infected themselves. Many bloodsucking insects, such as mosquitoes, ticks and fleas, are vectors. *(6 marks)*

Question 33 (Total 4 marks)

Natural or innate immune responses include general, non-specific defence mechanisms of plants and animals that are constantly present. These include passive physical and chemical barriers that prevent entry of pathogens. An active innate response in plants includes the hypersensitive response, which results in rapid cell death that reduces the spread of the pathogen or the formation of abscisic layers that cause diseased parts to break off. The inflammatory response and phagocytosis are examples of the innate immune response in animals.

Acquired immunity occurs in vertebrates as a specific response to pathogens, either as a result of exposure or vaccination. B cells and T cells are key parts of acquired immunity, as is the recognition of non–self-antigens, (surface chemicals on the pathogen) and the formation of specific antibodies that support the defence against the pathogen. *(4 marks)*

Question 34 (Total 8 marks)

The effectiveness of anti-pathogens can be measured in terms of reduction of mortality rates, duration or severity of the illness, or relapse rates when used as a treatment. A factor in this effectiveness is the development of pathogens that are resistant to pharmaceuticals. In some instances anti-pathogens are used as a preventative (e.g. before bowel surgery) and the effectiveness would be measured in a reduction of those contracting a particular disease. Consideration needs to be made as to whether the pathogen is generalist or specialist (i.e. targeting many species or a group of species of hosts, or only one species), and also whether the anti-pathogen is generalist or specialist. A final criteria in assessing effectiveness is the likelihood of experiencing side effects as a result of taking the anti-pathogen. Antibiotics commonly result in the side effect of diarrhoea.

Answer Q34 continues

Antibiotics have been widely used since the end of World War II. They act by killing bacteria. Initially they dramatically improved health and improved survival of bacterial infections and are still widely used. The World Health Organisation recently reported all countries are experiencing decreasing effectiveness of antibiotics due to the increasing resistance of bacteria (WHO Factsheet, 2018). No new classes of antibiotic have been developed in the last 25 years.

The discovery of new viruses is still occurring, with 50 new viruses being discovered since 1972. Viruses are non-cellular, so antivirals have often prevented the action of viruses by inhibiting some of their enzymes (as opposed to being able to actually destroy the virus). Research into the development of antivirals depended on the development of appropriate cell cultures that enabled testing of the effectiveness of various drugs on the viruses that were inoculated into the cell cultures. Some antivirals stimulate the action of the cytokine interferon that are part of the immune response to viruses. There is a very fine balance between the efficacy, toxicity and development of resistance in antiviral drugs. For example, the drug developed to treat the herpes virus (cold sores) is limited to topical applications because it is so toxic to cells. Combination therapy has been necessary for the treatment of such viruses as Hepatitis C. Response rates are still relatively low and relapse rates high in many antivirals. Development of virus resistance is also a limiting factor. The high cost of development, patents and the antiviral drugs themselves has greatly limited their effectiveness for treating diseases such as Hepatitis C. *(8 marks)*

Question 35 (Total 5 marks)

Dieback disease has impacted a number of Australian native plants, including the West Australian jarrah, *Eucalyptus marginata.* The cause is *Phytopthora cinnamomi*, a fungi-like soil-borne pathogen that attacks the roots, causing wilting, leaf-yellowing, root death and cankers on the trunk.

Investigation of secondary sources initially revealed information about preventative measures such as quarantine, which reduces the chances of soil movement, and treatment such as phosphite fungicide on the symptoms of the many native plants and fruit trees that can be impacted. The Royal Botanic Gardens Sydney focuses on disease management on its website.

Eventually a source revealed information about a research project that investigated the genomes of variants of both plants and the pathogen. Some plants were inoculated with the pathogen and the plant responses were observed. Responses included mass production of defensins and defence hormones such as abscisic acid. The plants also responded by lignin synthesis and rapid cell death near the site of pathogen entry (hypersensitive response). *(5 marks)*

CHAPTER 4

Module 8
Non-infectious Disease and Disorders

Part A Objective-response questions

1 The flowchart represents one example of homeostasis in an endotherm.

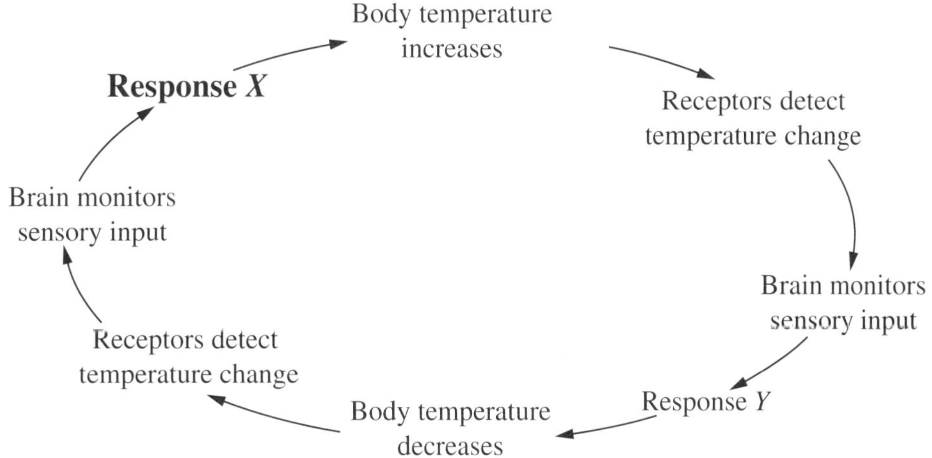

Which of the following does Response *X* represent in this cycle?

(A) Increased rate of sweat production

(B) Increased rate of urine production

(C) Decreased rate of sweat production

(D) Decreased rate of urine production

(q3, 2001 HSC)

2 In an endotherm, which of the following homeostatic responses would be produced by a sudden and prolonged decrease in ambient temperature?

(A) Decreased uptake of oxygen

(B) Decreased muscular activity

(C) Decreased blood flow to the skin surface

(D) Decreased rate of internal metabolic processes

(q2, 2002 HSC)

3 Spinifex is also called porcupine grass because its leaves can curl up into a needle shape.

The stomates are located in sunken grooves on the underside of the leaf and are enclosed as the leaf curls up.

Which process do these adaptations best reduce?

(A) Conduction

(B) Pollination

(C) Translocation

(D) Transpiration

(q1, 2003 HSC)

4 A small Australian mammal that lives in the alpine regions of New South Wales has specific features that enable it to retain body heat. Identify the features that are most likely to be present in the mammal described.

(A) Long ears, rounded body, long legs

(B) Short ears, rounded body, short legs

(C) Short ears, slender body, long legs

(D) Short ears, slender body, short legs

(q3, 2003 HSC)

5 The Australian hopping mouse, *Notomys alexis*, is a desert animal. It produces urine that is very concentrated.

Why is this an advantage for the animal?

(A) It needs to conserve water.

(B) It is nocturnal and only drinks at dusk.

(C) It has a high intake of salt in its specialised diet.

(D) It needs to excrete large amounts of water to survive.

(q1, 2001 HSC)

6 What is a role of the kidney in the excretory system of mammals?

(A) To remove salt from the body and to keep water in the body

(B) To remove water from the body and to keep salt in the body

(C) To remove nitrogenous waste from the body and to maintain water levels in the body

(D) To remove water from the body and to maintain levels of nitrogenous substances in the body

(q6, 2009 HSC)

7 Experiments were carried out on plants living in different environments to measure the size of the leaf stomata at different times of the day. Previous investigations had shown that plants transpire more water when the size of the stomata is larger.

Which graph best represents a plant living in a dry environment?

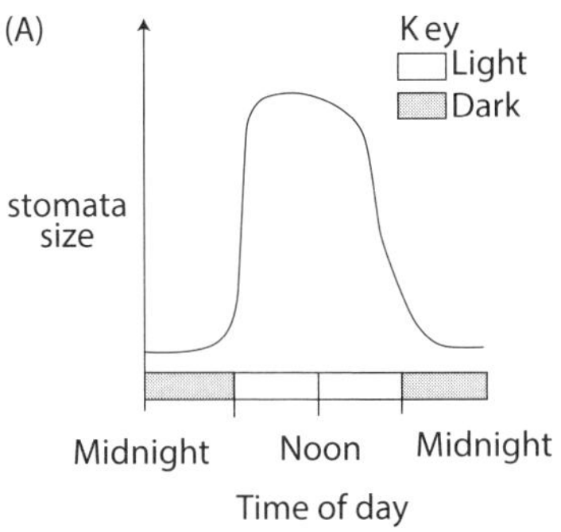

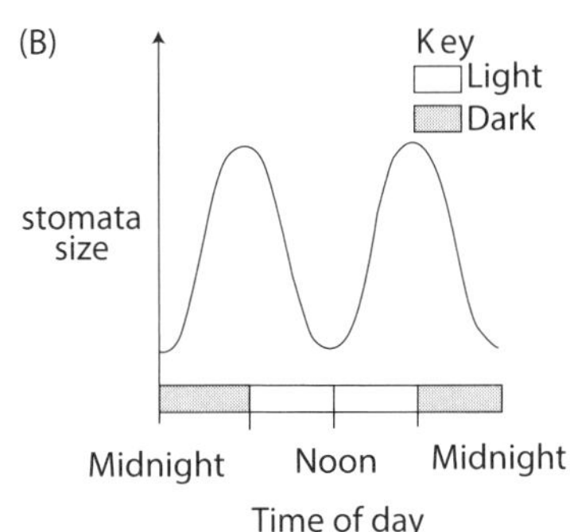

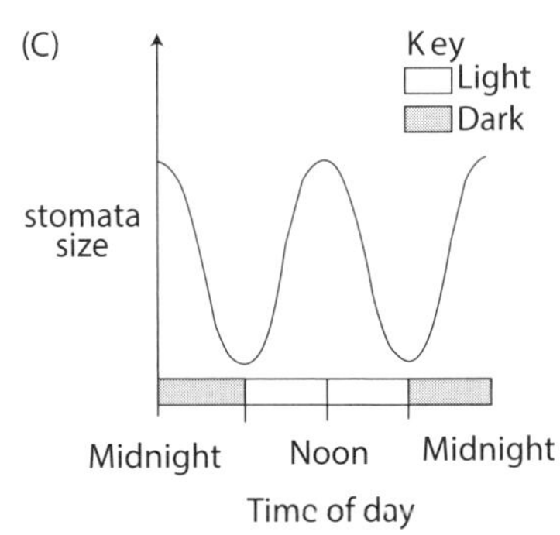

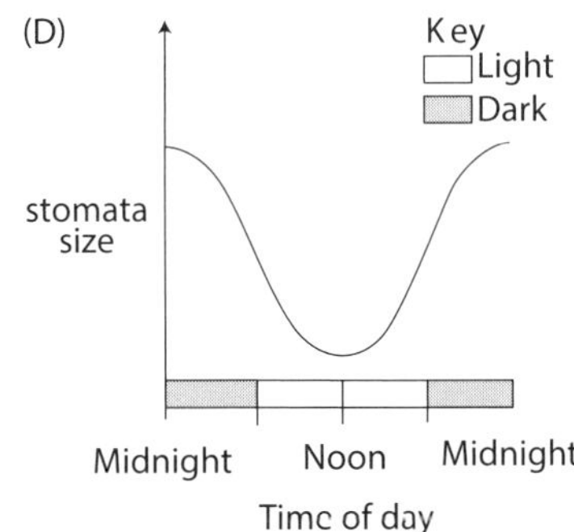

(q11, 2009 HSC)

8 Consider the diagram of the nephron.

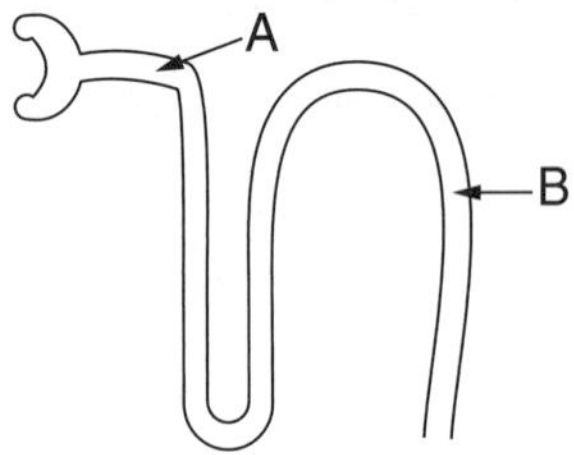

A diagram of a nephron

What happens to the concentrations of glucose, urea and protein as fluid moves from **A** to **B** in the nephron?

	Glucose Concentration	*Urea Concentration*	*Protein Concentration*
(A)	Unchanged	Decreases	Decreases
(B)	Unchanged	Decreases	Increases
(C)	Decreases	Increases	Increases
(D)	Decreases	Increases	Unchanged

(q8, 2010 HSC)

9 In organisms, the maintenance of a constant internal environment is:

(A) necessary because organisms must have a constant body temperature.

(B) necessary because enzyme activity is highest at specific temperatures.

(C) unnecessary because organisms are found in environments with a broad range of temperatures.

(D) unnecessary because the nervous system detects and responds to changes in ambient temperature.

(q3, 2011 HSC)

10

Huntington's Disease is caused by an inherited gene that codes for a toxic protein.

Kwashiorkor is a disease caused by a deficiency of proteins in the body.

Mesothelioma is a disease caused by a gene mutation in the lungs after exposure to asbestos.

Which row in the table correctly classifies these diseases?

	Huntington's Disease	*Kwashiorkor*	*Mesothelioma*
(A)	Genetic	Nutritional	Environmental
(B)	Nutritional	Environmental	Environmental
(C)	Genetic	Nutritional	Genetic
(D)	Nutritional	Environmental	Genetic

(q5, 2012 HSC)

11 Nitrogenous waste is at its highest concentration in:

(A) plasma in the renal vein.

(B) plasma in the renal artery.

(C) fluid in the collecting ducts of the kidney.

(D) interstitial fluid in the cortex of the kidney.

(q12, 2012 HSC)

12 Two experiments were conducted where either cold air or hot air was blown continuously onto a student's legs while the skin temperature on the student's arm was being measured.

The graph shows the change in skin temperature on the arm of the student for each experiment.

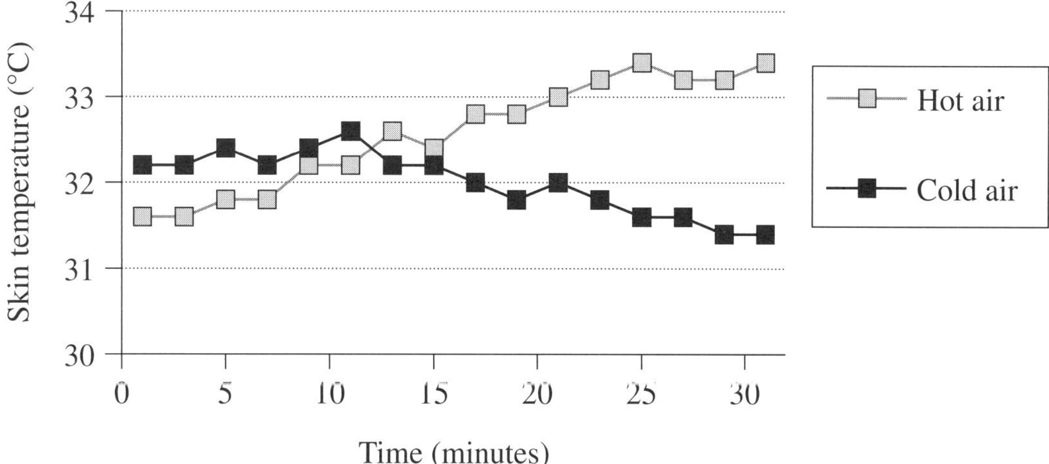

Which process best accounts for the trends shown in the graph?

(A) Diffusion

(B) Enantiostasis

(C) Homeostasis

(D) Inflammation

(q12, 2013 HSC)

13 Which statement defines homeostasis in multicellular organisms?

(A) Homeostasis is the process by which cells maintain their internal environment.

(B) Homeostasis is the maintenance of the internal and external environment of the organism.

(C) Homeostasis is the process by which animals and plants maintain their body temperature.

(D) Homeostasis is the maintenance of a constant internal environment of the organism.

(q9, 2004 HSC)

14 Lung cancer is one of the diseases linked to smoking.

To what category of disease does lung cancer belong?

(A) Environmental

(B) Infectious

(C) Nutritional

(D) Viral

(q3, 2005 HSC)

15 Which of the following best describes the main focus of epidemiology?

(A) Diseases related to the skin

(B) Changes in the characteristics of a species

(C) Factors involved in the occurrence, prevalence and spread of disease

(D) How the body maintains its functions in response to variations in the environment

(q6, 2005 HSC)

Part B Short-answer questions

Question 16 (8 marks)

The following diagram shows a rural coastal area and the associated towns, rivers and industry for each of the townships.

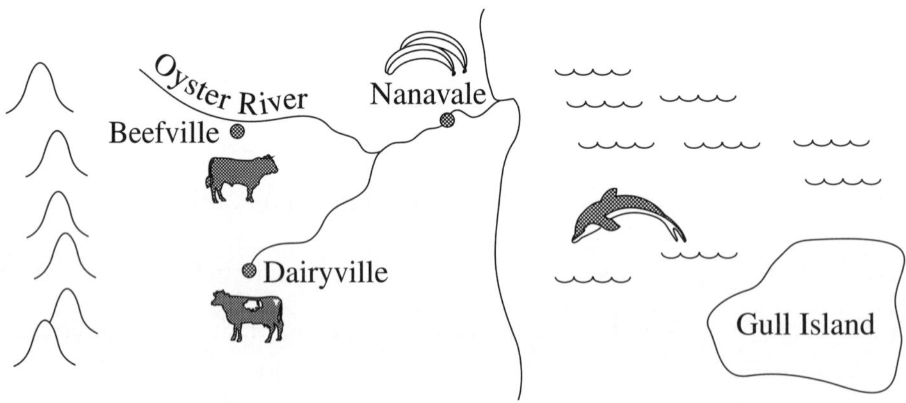

Question 16 continues

Question 16 (continued) **Marks**

An epidemic of a disease has broken out in Nanavale. The symptoms are stomach ache, vomiting and tiredness. Many families in Nanavale have only one member with the disease, therefore it is apparently non-infectious. The symptoms appear worse in infants than adults.

Isolated cases of this disease have occurred in the nearby towns of Dairyville and Beefville. No cases have been reported on Gull Island.

Design an epidemiological study to investigate the origin of the disease. **8**

(16 lines) (q29, 2003 HSC)

Question 17 (4 marks)

Draw a labelled diagram to show how a specific feedback mechanism plays an essential role in homeostasis. **4**

(8 line spaces) (q18, 2003)

Question 18 (6 marks)

(a) Draw the most appropriate graph for the tabled information, on the grid provided. **5**

Death due to lung cancer in males and females in Australia

	Death rate per 100 000 population	
Time	*Males*	*Females*
1994	59	19
1995	56	19
1996	55	19
1997	51	19
1998	52	18
1999	50	19
2000	48	19
2001	47	20

Question 18 continues

Question 18 (continued) **Marks**

Death due to lung cancer

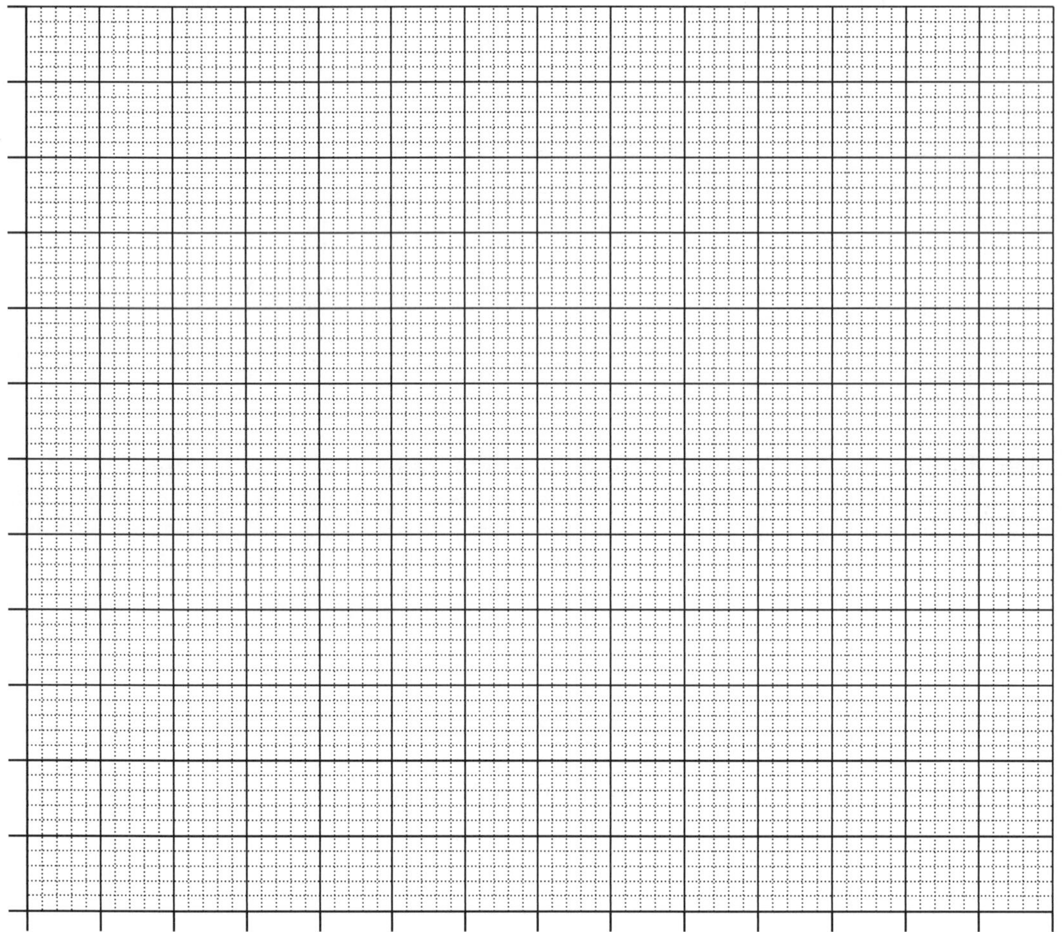

(b) Suggest additional data that needs to be gathered before a relationship between smoking and lung cancer can be inferred. **1**

(1 line) (q25, 2004 HSC)

Question 19 (5 marks)

(a) Name TWO refractive media in the human eye. **2**

(2 lines)

(b) Explain how a problem with ONE of these media might contribute to poor eyesight or blindness. **3**

(4 lines) (q28(a)(i) and (ii), 2007 HSC)

Question 20 (3 marks) **Marks**

Use an example to outline the role of the nervous system in maintaining homeostasis. **3**

(8 lines) (q17, 2005 HSC)

Question 21 (4 marks)

Epidemiological studies have demonstrated a relationship between ultraviolet radiation exposure and the development of melanoma, a type of skin cancer.

The graph demonstrates rates of melanoma in males and females between 1972 and 1997.

A student studying the graph made the following statement.

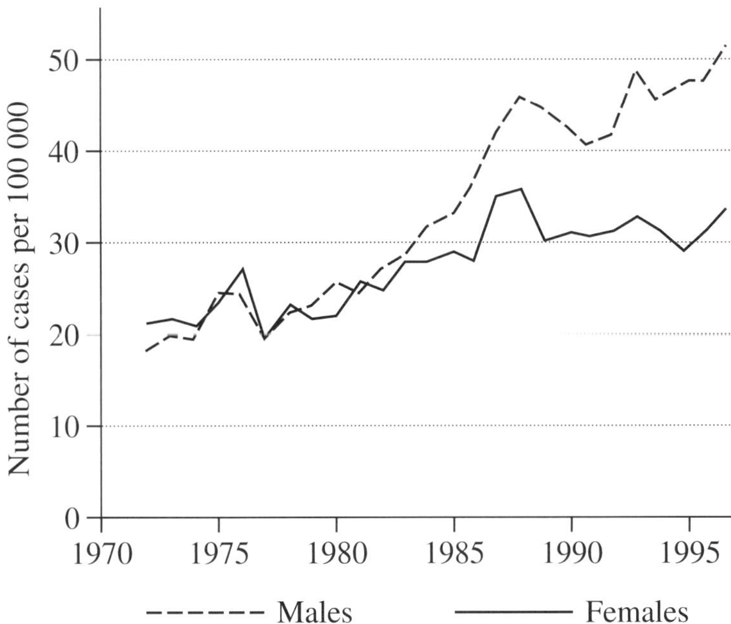

'The incidence of melanoma will continue to increase beyond 1997 at a greater rate in males than in females.'

Analyse the data in the graph to assess the validity of this statement. **4**

(8 lines) (q27, 2001 HSC)

Question 22 (7 marks)

Evaluate the appropriateness of TWO devices designed to assist people with different types of hearing impairment. **7**

(26 lines) (q29, 2001 HSC)

Question 23 (3 marks) **Marks**

Public health programs, pesticides and genetic engineering are used to control and/or prevent disease.

Using an example, explain how ONE of these strategies has been used to control or prevent disease within the community. 3

(6 lines) (q22, 2004 HSC)

Question 24 (6 marks)

The diagram represents a nephron, which is the functional unit of the kidney.

Nephrons make urine by:

- filtering small molecules and ions from the blood;
- reabsorbing the needed amounts of useful materials.

Surplus or waste molecules and ions flow out as urine.

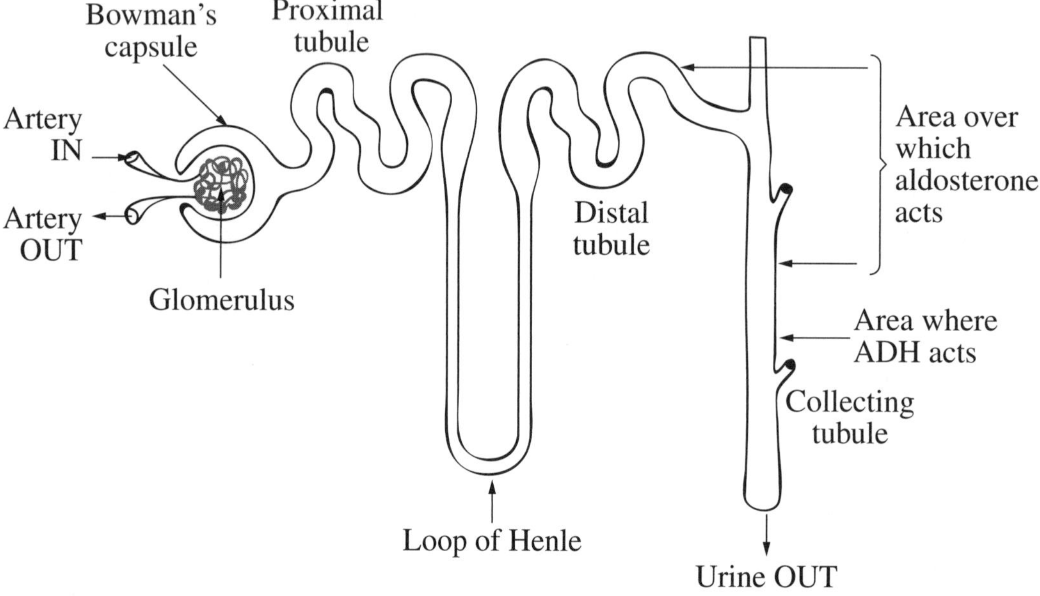

(a) Identify the area where filtration occurs, by marking it with an *X* on the diagram. 1

(b) Identify the area where reabsorption occurs, by shading it on the diagram. 1

(c) Discuss the importance of hormone replacement therapy for people who cannot secrete aldosterone. 4

(8 lines) (q23, 2002 HSC)

Question 25 (4 marks) **Marks**

Consider the following diagram.

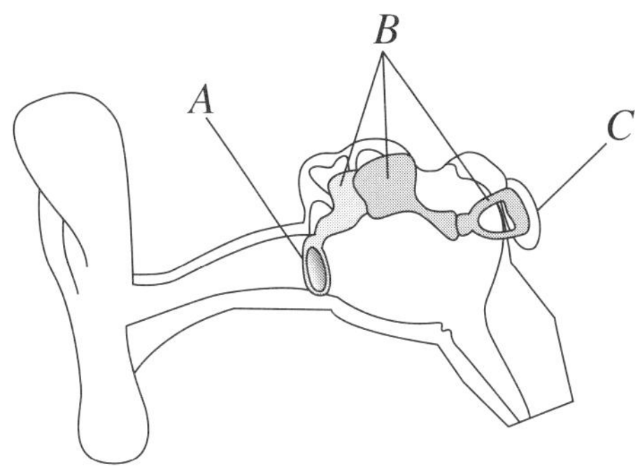

(a) Identify the parts *A*, *B* and *C*. 2

(2 lines)

(b) Outline the functions of part *B*. 2

(4 lines) (q31(a)(i) and (ii), 2012 HSC)

Question 26 (5 marks)

A scientist performed an epidemiological study to investigate the cause-and-effect relationship of smoking and lung cancer as follows:

1. Handed out a scientifically valid questionnaire to all colleagues (n = 144) at work.
2. Checked that there was an equal number of male and female respondents.
3. Discovered there were more non-smoking respondents than smoking respondents. Removed some of the non-smokers until both groups had equal numbers.
4. Checked that all respondents had a medical check-up in the past year.
5. Analysed data, wrote the paper and published it in a scientific blog.

From the information provided, analyse the methodology used by this scientist. 5

(10 lines) (q26, 2012 HSC)

Question 27 (4 marks)

Explain how the differences between photoreceptor cells in the eye contribute to human vision. 4

(10 lines) (q31(c), 2014 HSC)

Question 28 (7 marks) **Marks**

(a) Outline ONE essential step that must occur for the photoreceptor signal to be transmitted along the optic nerve to the brain. 1

(1 line)

(b) Microelectronic chips containing light-sensitive detectors have been implanted in the retinas of blind volunteer patients. These implants do not give normal vision but allow these patients to see only areas of light and dark.

Relate your understanding of the use of hearing aids and cochlear implants to possible advantages and limitations of microelectronic chips to vision. 6

(15 lines) (q31(e), 2013 HSC)

Question 29 (5 marks)

(a) Describe the process of accommodation. You may include diagrams in your answer. 4

(6 lines)

(b) Outline how ONE technology is used to treat hyperopia. 1

(2 lines) (q31(c)(i) and (ii), 2013 HSC)

Question 30 (4 marks)

Lactose intolerance is a non-infectious disease found in some humans who cannot digest milk. The causes of non-infectious disease are grouped into three categories. An epidemiological study was undertaken to investigate the cause of lactose intolerance.

The study found variation in the occurrence of lactose intolerance in human populations from different parts of the world. Some of the data are shown in the table.

Human population	*Lactose Intolerance (%)*
Thai	99
Yoruba (Nigeria)	99
Fuloni (Nigeria)	18
Swiss	12
Swedish	2

Using information from the table, outline how the study could be continued to determine the cause of the disease. 4

(8 lines) (q27, 2006 HSC)

Question 31 (4 marks) **Marks**

Explain the relationship between the cause and ONE symptom of ONE named non-infectious disease. **4**

(8 lines) (q24, 2001 HSC)

Question 32 (4 marks)

Investigators gathered data on a group of 100 smokers for a period of 10 years.

During this time, 12 people in the group developed lung cancer, two died in traffic accidents, and three died of heart attacks. The investigators used this data to state that smoking caused lung cancer.

Describe how this investigation could be improved. **4**

(8 lines) (q18, 2002 HSC)

Question 33 (5 marks)

Discuss the role of public education in preventing cancers. **5**

(12 lines) (q31(c), 2001 HSC)

Question 34 (6 marks)

Evaluate the effectiveness of genetic engineering used to treat non-infectious disease. **6**

(18 lines) (Sample question)

Question 35 (4 marks)

Describe mechanisms in plants that help to maintain water balance. **4**

(15 lines) (Sample question)

Module 8
Non-infectious Disease and Disorders

Answers

Part A Objective-response questions

1 C Response *Y* causes the body to increase sweat production, enabling it to cool down. Once sufficiently cool, the body needs to stop excess perspiration and Response *X* decreases sweat production.

2 C Blood is directed away from the extremities to retain core body temperature.

3 D Water loss occurs through leaves by the process of transpiration.

4 B The surface–area-to-volume ratio needs to be small to reduce heat loss.

5 A Concentrated urine reduces water loss in the desert-dwelling endotherm.

6 C Mammalian kidneys remove toxic nitrogenous waste (urea) and help maintain water and salt balance.

7 D This is the only graph to show consistent decrease in stomata size as temperature increases during daylight.

8 D The concentrations of proteins will remain unchanged. Glucose will be reabsorbed and the urea will increase as the blood passes through a nephron.

9 B Enzyme function is dependent on optimum temperature being maintained constantly.

10 A Inherited characteristics are genetic. Protein deficiency can be caused by poor diet, so it is nutritional. Mutations caused by exposure to asbestos are environmental.

11 C The waste is taken from the collecting ducts to the bladder for removal from the body.

12 C The temperature fluctuation on the arm correlates with the temperature of the air being blown onto the legs, reflected by vasoconstriction/dilation in order to maintain body temperature, which is a homeostatic function.

13 D Homeostasis is involved with the maintenance of *constant* internal conditions.

14 A Lung cancer belongs to the environmental category, because it is a direct result of an individual's activity and not associated with infectious or communicable diseases. It is not nutritional, as tobacco is not considered to be a food.

15 C Epidemiology is the study of disease in a population, so C is the only relevant option.

Part B Short-answer questions

Question 16 (Total 8 marks)

Epidemiological studies are used to study diseases that affect large numbers of people in a population. The key features of an epidemiological study include the collection of large amounts of relevant data, statistical analysis of the information generated and continued monitoring of the population over an extended length of time. The information is then used to isolate the cause of a disease and develop a treatment/control measure. *(3 marks)*

To determine the cause of the disease in the location depicted the following steps should be carried out:

1. Interview all persons affected by the disease. Information about their age, sex, diet, occupation, lifestyle and general movements in the area will need to be gathered.
2. Interview a similar sample of people in the area who have not been affected by the disease. These will need to be compared to communities that are not directly associated with the area such as Gull Island. Both surveys should include a large number of people and include the collection of the same data gathered from the disease-affected group.
3. The data need to be analysed statistically to identify any patterns or trends and identify similarities and differences between the non-affected groups and compare these with the affected groups.
4. A possible cause for the disease will be identified and a management plan developed to prevent the disease. This may include various tests of possible cures or prevention strategies and education of the populations and the development of management strategies for the environment.
5. The population will need to be monitored periodically. *(5 marks)*

Question 17 (Total 4 marks)

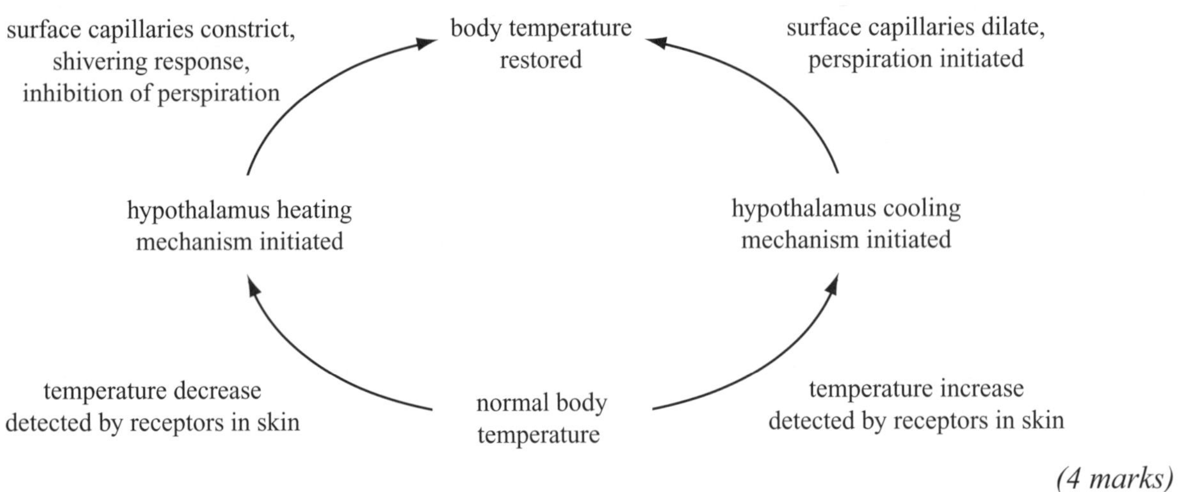

(4 marks)

Question 18 (Total 6 marks)

(a)

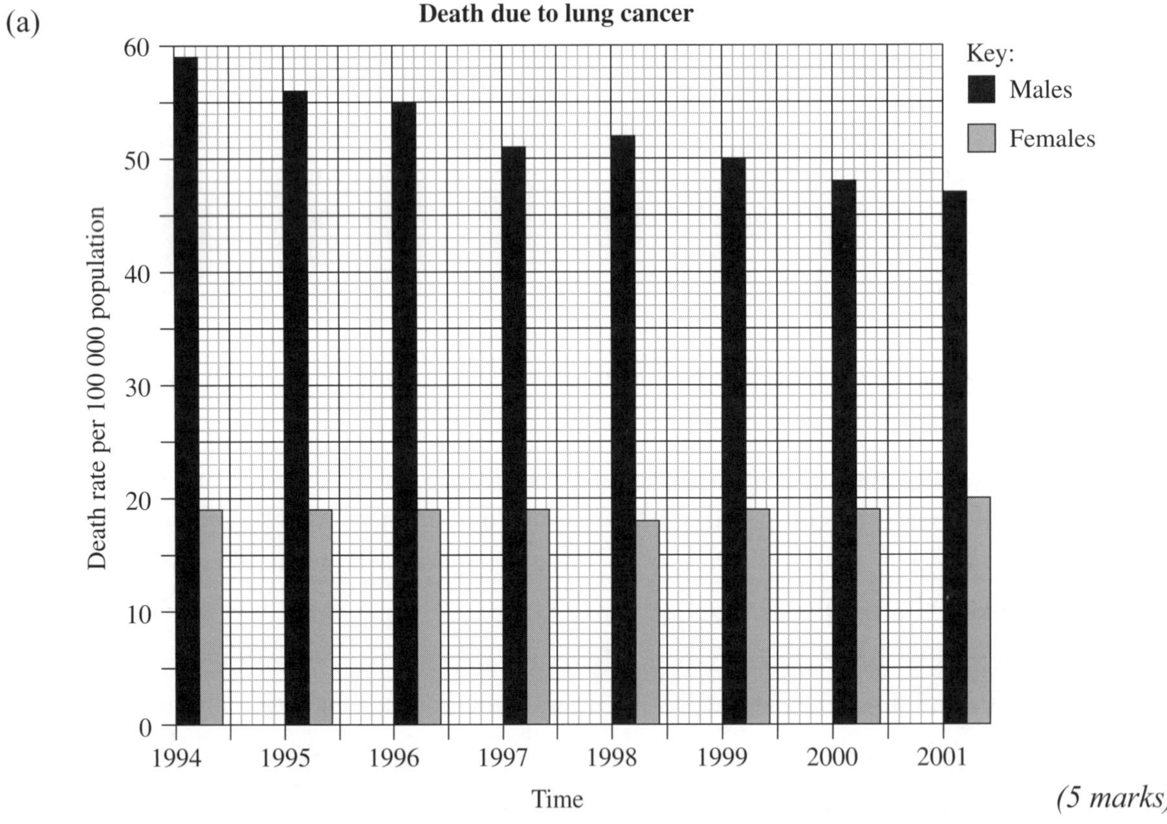

(5 marks)

(b) Before a relationship between smoking and lung cancer can be inferred, additional information such as the duration individuals smoked for, or the quantity of cigarettes smoked by individuals, would be required. *(1 mark)*

Question 19 (Total 5 marks)

(a) Refractive media in the eye include: cornea, aqueous humour, lens, vitreous humour. *(2 marks)*

(b) (Many problems can exist with these refractive media, so answers will be very varied.)

Lens (example): A problem with the lens can result in poor eyesight or even blindness. A lens may become cloudy due to cataracts forming within it, resulting in loss of transparency, thereby causing diminished vision or even total blindness, depending on the severity of the growth. A cataract can be caused by aging, but it can be present from birth, or it can develop from disease such as diabetes, use of drugs such as corticosteroids, or injury such as trauma or exposure to excessive infrared radiation.

Other problems with the lens include the inability to change the shape/thickness of the lens, resulting in hyperopia/myopia. *(3 marks)*

Question 20 (Total 3 marks)

(Possible answers include the nervous system used to maintain the internal temperature, blood sugar, or reflex actions.)

Example: To maintain a constant internal temperature, the body will use a feedback mechanism to detect an increase/decrease in external temperature. This change will be passed on from the receptor organ to the CNS, and the brain will respond by sending a message to the effector organ, such as a muscle or gland, to shiver or perspire. Once the required temperature has been restored, the negative feedback will reverse this message to prevent overcorrection. *(3 marks)*

Question 21 (Total 4 marks)

The graph depicts an increase in the incidence of melanoma, although the rate of increase is generally greater for males than for females between 1983 and 1995. The statement predicts that this trend will continue; however, it is difficult to state this conclusively as other variables that may affect this pattern have not been identified and only one State has been considered. Two factors not considered in this graphical representation include the location and age of the participants, factors that would have considerable impact on the validity of such a statement. *(4 marks)*

Question 22 (Total 7 marks)

(The answer should include two devices, clearly identified. As an evaluation is sought it is important to include both advantages and limitations of the device and its application. Examples could include cochlear implants, various types of hearing aids, ossicle replacement and telephone amplification devices.)

Example: Cochlear implants are used to replace the damaged cochlea. The cochlea normally is responsible for the transmission of different sound frequencies to the auditory nerve. If the cochlea is damaged, amplified hearing aids do not work and so implants can be used to overcome inner ear problems. Some limitations of using cochlear implants can include the inability to detect all frequencies, resulting in sound distortion. The recipient is required to wear a permanent device attached to the outside of the skull, which is a hindrance. *(7 marks)*

Question 23 (Total 3 marks)

Public health programs are used to educate wider general populations regarding the dangers of particular diseases or practices, enabling people to identify symptoms and take preventative action to reduce the incidence. Programs use advertisements and pamphlets.

One such program includes the 'Slip, slop, slap' campaign used to raise awareness of the danger of exposure to harmful UV radiation from the Sun. By educating the population about the dangers of excess exposure to sunlight without protection such as clothing, hat or sunscreen, the incidence of cancers such as melanoma can be reduced in future generations. Together with regular screening and early intervention, the incidence of skin cancer may be reduced.

Examples of other programs: AIDS and sexually transmitted diseases campaigns, early childhood immunisation, 'Life. Be in it.' campaign. *(3 marks)*

Question 24 (Total 6 marks)

(a) and (b)

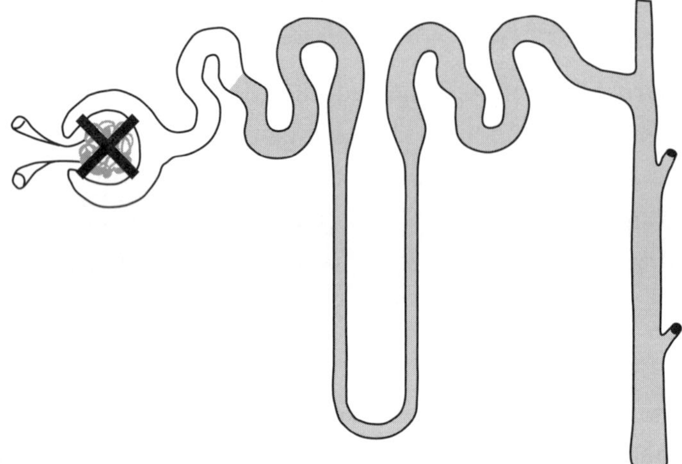

(2 marks)

(c) Hormone replacement therapy (HRT) is used by patients who cannot secrete for themselves hormones required by the body. The role of aldosterone in the nephron is to control the permeability of the tubule to regulate the levels of sodium. This results in water reabsorption, ensuring that the water, and hence blood pressure, is maintained constantly. In the case of insufficient aldosterone secretion by the body, HRT can be used. This will enable the blood pressure to stabilise, thereby enabling appropriate functioning of the nephron by promoting reabsorption. *(4 marks)*

Question 25 (Total 4 marks)

(a) *A*, ear drum/tympanic membrane; *B*, ossicles; *C*, oval window *(2 marks)*

(b) The middle ear ossicles (made up of hammer, anvil and stirrup) transmit sound vibrations through the middle ear. Vibrating air molecules cause the tympanic membrane to vibrate, making contact with the hammer. The three small bones that make up the ossicles act as levers to magnify and transmit the vibration so that is received by the oval window. This sets up pressure changes that vibrate the fluid in the inner ear. From here, the vibrations are received by the organ of corti within the cochlea in the inner ear, which are connected to the auditory nerve. *(2 marks)*

Question 26 (Total 5 marks)

The methodology described will provide general results only and would not be considered valid in terms of the number studied (144). Further, the scientist removed non-smoking subjects to even out the numbers of smokers and non-smokers, thereby reducing the numbers even more. There was no indication of the basis on which the scientist selected the subjects that were eliminated from the study other than the fact that there were more non-smokers than smokers. In order for the study to be valid, the larger sample size without bias being introduced is essential. Removing some of the non-smokers can skew the results by removing links to other possible sources for the cancers. Checking whether the respondents had medical checks in the past year may only identify that the symptoms were not present and not accurately identify whether the subjects had any changes that were not exhibited yet. You would require specific tests on the lungs themselves to determine this accurately. Other useful information that needs to be included about smoking habits and history of the participants includes history about living/working/socialising with smokers.

Without such information, the scientist's conclusions would not be very significant in terms of medical findings. *(5 marks)*

Question 27 (Total 4 marks)

The eye has two types of photoreceptor cells located within the retina. The rods and cones, which both contain light-sensitive pigment, contribute differently to human vision and work together to produce the clear, sharp images that are possible by the human eye. The following table summarises the differences between the photosensitive cells in the eye.

Feature	*Cones*	*Rods*
Distribution	Cones are found in the middle of the retina, concentrated at the fovea region.	Rods are found across the retina but are most numerous around the periphery. They are absent in the fovea.
Structure	These cone-shaped cells contain stacks of membranes where the photosensitive photopsin pigment is found. There are three types: red, blue and green.	These rod-shaped cells contain stacks of membranes where the photosensitive rhodopsin pigment is found.
Function	Each of the three cone types is sensitive to different light intensity. This produces colour vision but relies on the stimulation of opsin by visible light.	Rhodopsin is sensitive to blue green light and works in low or dim light. Rods detect movement and shape but do not assist with detail vision, because they have only one pigment. They can only tell light from dark.

(4 marks)

Question 28 (Total 7 marks)

(a) In order for the photoreceptor signal to be transmitted to the brain, neuron stimulation must reach a threshold. If the threshold is reached, the signal is transmitted. If it is not reached, there is no signal. This is known as an all-or-nothing response. *(1 mark)*

(b) Hearing aids, and more recently cochlear implants, are used to augment hearing.

A hearing aid is an external device that sits on or just inside the outer ear. It amplifies mechanical vibrations so that they will pass into the middle ear. The wearer will be able to hear only if the cochlea still has some sensitivity and the auditory nerve still functions. Hearing aids do not restore normal hearing and cannot eliminate background noise, so the wearer needs to adjust to interpreting sounds under different conditions.

A cochlear implant is an electronic device that provides a sense of sound to profoundly deaf patients. It works by bypassing the damaged parts of the inner ear and directly stimulating the auditory nerve. Each implant consists of two parts. One part, containing electrode wires that are inserted into the cochlea, is surgically implanted in the skull. This is connected to a second, external component consisting of a microphone with a speech processor that converts sound into electrical impulses. Cochlear implants do not restore normal hearing, and recipients must learn to interpret the signals the implant creates.

This has to be combined with techniques such as lip reading to truly understand conversation. However, this is better than no hearing at all.

Microelectronic chips containing light-sensitive detectors would be similar to cochlear implants in that they would not restore complete vision, but would enable the recipient to detect changes in light. This would be a significant improvement if the recipient was completely blind. The ability to detect shadows or moving objects, together with the use of a cane, guide dog and braille, would make it easier for the recipient to move around more confidently. A limitation of microelectronic chips is that the signals received may appear as shadows, possibly pixelated, on a white-to-black spectrum. As is the case with cochlear implant recipients, interpretation of these signals/images would take some getting used to.

(6 marks for applying an understanding of how hearing aids and cochlear implants assist hearing-impaired people compared with how microelectronic chips work to assist sight-impaired people; remember to include advantages and limitations)

Question 29 (Total 5 marks)

(a) Accommodation is the process by which the lens of the eye changes shape to focus an image onto the retina, irrespective of the distance of the object away from the eye.

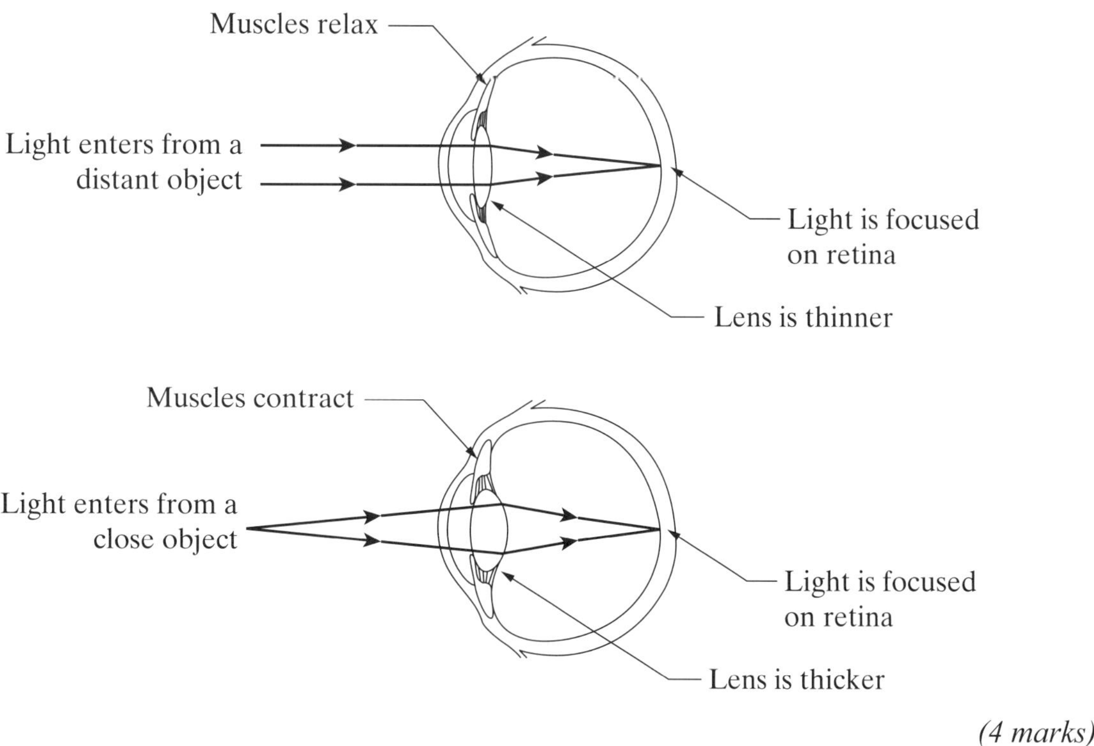

(4 marks)

(b) Hyperopia, or long-sighted vision, occurs when the eyeball is too short or the lens cannot become round enough to focus the object onto the retina. This makes it difficult to focus on close objects, but can be rectified using a convex lens.

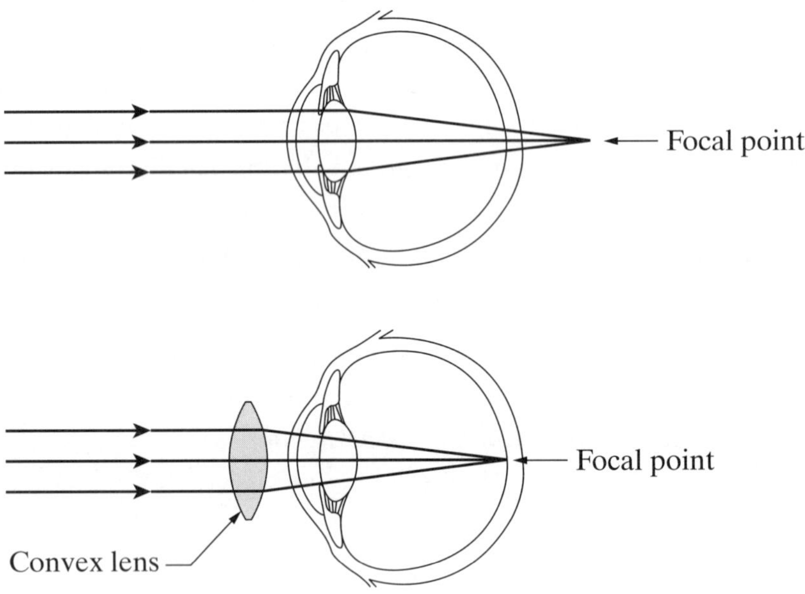

(1 mark)

(Please note: diagrams may be used in the answer to this question, but are not essential for full marks.)

Question 30 (Total 4 marks)

The data provided has identified the number of each population affected by the disease.

In order to continue the study to determine the possible cause of the lactose intolerance, studies can be carried out on the following:

- the specific diets of the individuals represented by each of the groups;
- the source of the lactose they are ingesting/drinking (e.g. cow, goat);
- the quantities that they are ingesting/drinking.

In undertaking further epidemiological study, attention needs to be paid to:

- the occurrence and transmission of the disease in the population and the environment;
- consideration of the risk factors to study participants;
- control of the disease, prevention of the disease, and possible elimination of the disease.

(4 marks)

Question 31 (Total 4 marks)

Example: Scurvy is a non-infectious disease caused by a deficiency of Vitamin C.

It results in poor wound healing, fragile capillaries, and sore or bleeding gums. A lack of dietary Vitamin C results in these symptoms. *(4 marks)*

Question 32 (Total 4 marks)

The investigation could be improved by increasing the sample size, and then repeating the study using other groups of people with similar characteristics but in different locations or with differing levels of age, health or fitness.

The use of a control group of non-smokers in the area would also serve to verify results. The data would need to be analysed statistically to verify any trends or patterns in the study groups.
(4 marks)

Question 33 (Total 5 marks)

The role of public education is very important in the prevention of cancers. Public education includes raising people's awareness of the risks involved in Sun exposure, preventative measures that could be taken to minimise risk, early detection and intervention options available and treatment for symptoms. Education programs may reduce the future incidence of cancer and enable those for whom it is too late to avoid the exposure to adopt appropriate management techniques such as the awareness of early signs and access to detection and treatment in order to prevent large-scale surgery. If signs and symptoms go undetected the cancers can spread, leading to prolonged treatment, surgery and unnecessary suffering, all of which can add a huge burden to the cost of public health.

Cancer is often a long-term disease so it is difficult to establish the effectiveness of campaigns because there is a time lag between a public campaign and the development of symptoms. For example, sunbathers can develop symptoms from their environmental past time many years later, so the effectiveness of any kind of 'Slip, slop, slap' campaign today might not be known for decades.

(5 marks for elaborating on a number of ways that public education can help the prevention or early detection and treatment of cancer and the difficulties in providing evidence of the effectiveness of campaigns because of time delays)

Question 34 (Total 6 marks)

Genetic engineering (GE) has been successfully used to produce animal models for research into treatments of diseases such as cancer, motor neurone disease, Parkinson's disease and Alzheimer's disease.

GE has also been used to produce hormones such as insulin for treatment of autoimmune diseases such as diabetes.

GE can be used to introduce corrected genes into a patient's stem cells and these stem cells then introduced back into the patient with a genetic disorder. This process is called gene therapy and examples are as follows:

- Research progress has been slow because of the difficulty in finding safe and effective vectors to carry the corrected genes into the somatic cells of patients. Severe combined immune deficiency (SCIDs) was one of the first conditions to have trials of gene therapy. Initial success was followed by disappointment when some patients suffered leukaemia, triggered by the viral vector. Successful treatment using gene therapy has occurred with patients suffering some forms of degenerative blindness and others with adenosine deaminase deficiency. The severity of bleeds in some haemophiliacs has also been reduced by gene therapy. Lipoprotein lipase (LPL) gene has been successfully introduced into muscle cells to prevent the toxic build-up of fat in the blood of patients. Cystic fibrosis shows potential as the gene delivery to the lungs could occur through aerosols but this requires further research.

- The inherited blood disorder beta thalassemia has been successfully treated using gene therapy. There is potential for sickle cell anaemia to also be treated using gene therapy.
- T-VEC is a treatment based on modifying the herpes simplex 1 virus so that it does not cause cold sores but instead kills cancer cells without impacting normal cells, as well as helping to activate a patient's immune cells to also recognise and fight cancer cells. This proved effective against late-stage melanoma.
- Some leukaemia sufferers had their immune cells removed, genetically altered and returned to effectively (in about 50% of cases) go into remission. *(6 marks)*

Question 35 (Total 4 marks)

Terrestrial plants generally absorb water through the extensive surface area of the root hairs and translocate it up to the leaves through the xylem. Water is also produced in cellular respiration. Water leaves the plant via the stomata in leaves that are controlled by guard cells. Water is used in photosynthesis; to keep cells turgid; as a solvent and the main component of cytoplasm; to translocate the sugar products of photosynthesis in solution; and to cool the plant through transpiration. At times the uses of water are in competition or a water shortage may override some functions of water in the plant. As both gaseous exchange and transpiration occur through the stomata, the plant may experience a 'conflict' between the need for carbon dioxide for photosynthesis and the need to reduce transpiration in dry conditions.

Plants in dry environments exhibit a number of responses and adaptations that help to maintain a water balance by reducing water loss or improving the ability to absorb water through the deep or extensive root systems.

Adaptations include heat reflection through size, shape and silvery colour of the leaf surface, rolled leaves or leaf hairs to retain humidity near stomata and a waxy cuticle to reduce evaporation from leaf surfaces. Some plants have water storage that holds the products of the light reaction of photosynthesis that must occur during the day until night time when stomata open to allow carbon dioxide to complete the formation of glucose. Cladodes and phyllodes are structures that are green and often look like leaves but have a lower density of stomata for the same surface area that absorbs light for photosynthesis. *(4 marks)*

CHAPTER 5

Sample
HSC Examination Paper
with 2015 HSC Questions

Biology

General Instructions

- Reading time — 5 minutes
- Working time — 3 hours
- Write using black pen
- Draw diagrams using pencil
- NESA approved calculators may be used
- For questions in Section II, show all relevant working in questions involving calculations

Total marks: 100

Section I – 20 marks
- Attempt Questions 1–20
- Allow about 35 minutes for this section

Section II – 80 marks
- Attempt Questions 21–36
- Allow about 2 hours and 25 minutes for this section

Section I

20 marks
Attempt Questions 1–20
Allow about 35 minutes for this section

---

1 Which statement below is the correct statement?

(A) All plants and animals have evolved innate immune responses to respond to and prevent entry of pathogens.

(B) Only plants have evolved innate immune responses to respond to and prevent entry of pathogens

(C) All plants and animals have evolved innate and adaptive immune responses to respond to and prevent entry of pathogens

(D) All animals have evolved an adaptive immune response to respond to and prevent entry of pathogens

(Sample question)

2 What are alternative forms of genes called?

(A) Alleles

(B) Autosomes

(C) Chromatids

(D) Chromosomes

(2015 HSC)

3 Which adaptation would decrease water loss from a plant in a region with low rainfall?

(A) Broad leaves

(B) Surface roots

(C) Sunken stomates

(D) Loosely packed epidermal cells

(2015 HSC)

4 Which of the following statements best explains how reproduction ensures the continuity of species?

(A) Continuity of a species occurs through diversity.

(B) Continuity of a species occurs through asexual reproduction and the replication of genetic material.

(C) Continuity of a species occurs through sexual reproduction, the replication of genetic material (DNA) and its transfer to the next generation.

(D) Continuity of a species occurs through both asexual or sexual reproduction, the replication of genetic material (DNA) and its transfer to the next generation.

(Sample question)

5 The diagram shows a model of some types of chromosomal rearrangements.

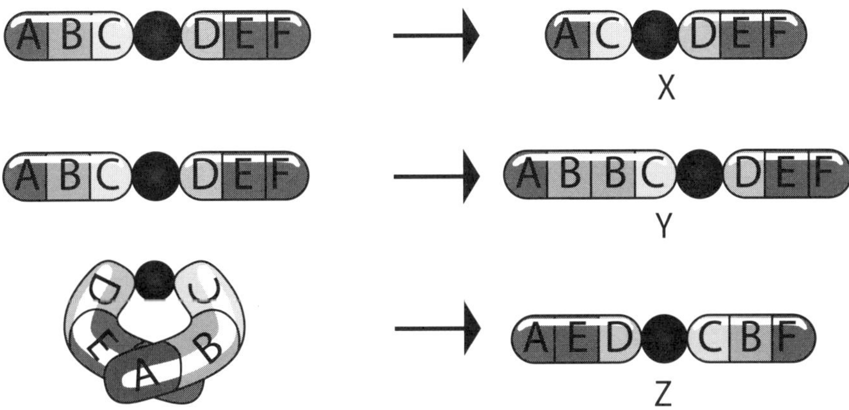

What type of rearrangement is represented by Y?

(A) Inversion

(B) Deletion

(C) Inversion and duplication

(D) Duplication

(Sample question)

6 What is the term given to disorders caused by organisms such as bacteria, fungi, viruses or parasites?

(A) Pathogen

(B) Infectious disease

(C) Non-infectious disease

(D) Contagious disease

(Sample question)

Refer to the following information to answer Questions 7 and 8.

The diagram shows a homeostatic mechanism in a mammal.

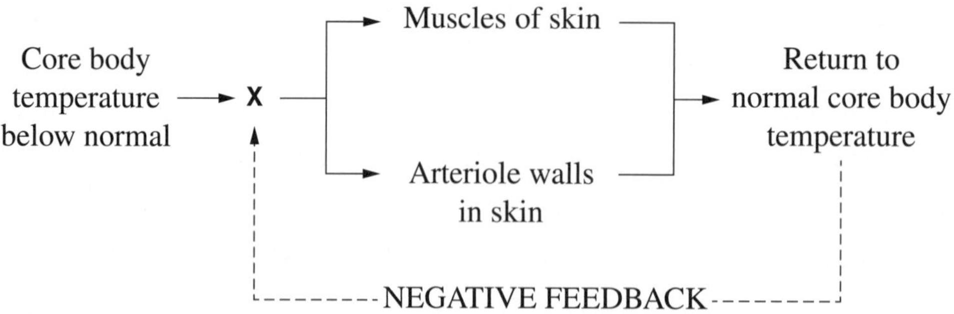

7 What does **X** represent in the diagram?

(A) The heart

(B) The brain

(C) A thermoreceptor in the skin

(D) A pressure receptor in a blood vessel

(2015 HSC)

8 Which of the following describes what happens to the muscles and the arteriole walls in the skin when the core body temperature is below normal?

	Muscles of skin	*Arteriole walls in skin*
(A)	Relax to lower epidermal hairs	Expand
(B)	Contract to raise epidermal hairs	Contract
(C)	Relax to raise epidermal hairs	Expand
(D)	Contract to lower epidermal hairs	Contract

(2015 HSC)

9 A pathogen and a red blood cell are drawn to the same scale, with some features indicated.

Pathogen

Cell wall

Membrane

Red blood cell

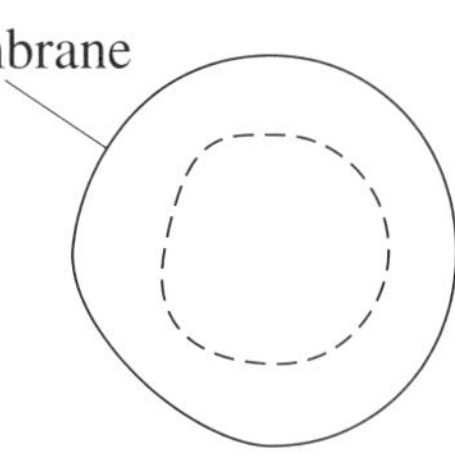

What type of pathogen is this?

(A) A virus

(B) A prion

(C) A fungus

(D) A bacterium

(2015 HSC)

10 What are somatic cells?

(A) Normal body cells in prokaryotes containing chromosomes in pairs

(B) Normal body cells in eukaryotes containing chromosomes in pairs

(C) Normal sex cells in prokaryotes

(D) Normal sex cells in eukaryotes

(Sample question)

11 Environment can affect phenotype by altering the sequence of bases in DNA.

Which of the following is an example of this?

(A) High protein diets making children taller than their parents

(B) Stress causing the expression of one set of genes instead of another

(C) Language and music lessons improving intelligence in young children

(D) Nuclear fallout from atomic bombs increasing birth defects in populations

(2015 HSC)

Refer to the following information to answer Questions 12 and 13.

The larvae of fruit flies damage fruit in Western Australia. To control the problem, growers are advised to spray fruit trees with pesticides. Any already damaged fruit is boiled and disposed of as chicken food or landfill. Another control measure is the release of genetically engineered infertile flies of this species.

12 Which reason best explains why the corresponding control measure reduces this problem?

	Control measure	*Reason*
(A)	Release of infertile flies	Infertile flies do not eat the fruit
(B)	Release of infertile flies	The number of flies in the next generation is decreased
(C)	Boiling fruit and feeding to chickens	Chickens are unaffected by the damaged fruit
(D)	Boiling fruit and feeding to chickens	The amount of waste for landfill is reduced

(2015 HSC)

13 The following measures could be used to prevent the spread of this fruit fly across Australia.

1. Australia-wide release of infertile fruit flies
2. Aerial spraying of orchards throughout the country
3. Spot spraying of newly affected orchards in Western Australia
4. Stopping the transport of fruit from Western Australia to other states

To prevent the spread of this fruit fly across Australia, which combination of measures would be most practical to use?

(A) 1 and 2

(B) 1 and 4

(C) 2 and 3

(D) 3 and 4

(2015 HSC)

14 The table shows the base triplets in mRNA for amino acids.

From the table, the amino acid Serine (Ser) can be coded for by the base triplet UCG.

Base triplets found in messenger RNA

First base	Second base U	Second base C	Second base A	Second base G	Third base
U	Phe	Ser	Tyr	Cys	U
U	Phe	Ser	Tyr	Cys	C
U	Phe	Ser	Stop	Stop	A
U	Phe	Ser	Stop	Trp	G
C	Leu	Pro	His	Arg	U
C	Leu	Pro	His	Arg	C
C	Leu	Pro	Gln	Arg	A
C	Leu	Pro	Gln	Arg	G
A	lle	Thr	Asn	Ser	U
A	lle	Thr	Asn	Ser	C
A	lle	Thr	Lys	Arg	A
A	Met	Thr	Lys	Arg	G
G	Val	Ala	Asp	Gly	U
G	Val	Ala	Asp	Gly	C
G	Val	Ala	Glu	Gly	A
G	Val	Ala	Glu	Gly	G

Which base triplet could code for the amino acid Tyrosine (Tyr)?

(A) CCU

(B) CAU

(C) UAA

(D) UAC

(2015 HSC)

15 In a certain plant species, individual plants have either yellow, red or orange flowers.

Two plants, each with a different flower colour, were crossed in a breeding experiment like those carried out by Mendel. The F2 results were: 6 red, 11 orange and 5 yellow flowered plants.

What were the genotypes of the original parent plants?

(A) RY and RY

(B) RR and rr

(C) RR and YY

(D) Rr and RY

(2015 HSC)

16 Why might epidemiology be considered more essential for the study of non-infectious diseases than for the study of infectious diseases?

(A) The causes of infectious diseases have already been determined.

(B) Only non-infectious diseases are affected by patterns of behaviour.

(C) Epidemiology cannot be used to find the causes of infectious diseases.

(D) Koch's postulates are not useful in finding the causes of non-infectious diseases.

(2015 HSC)

17 The diagram shows an interaction between cells of the immune system.

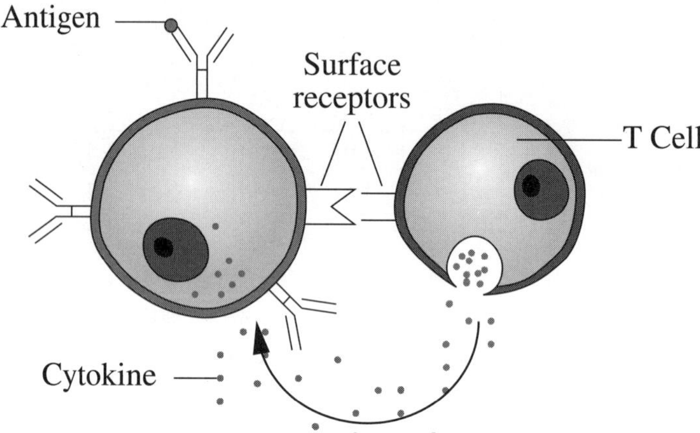

What specific process is shown in the diagram?

(A) B cell encountering an antigen

(B) Activation of a macrophage by a helper T cell

(C) Stimulation of a B cell to become a plasma cell

(D) Cytotoxic T cell destroying a virus-infected cell

(2015 HSC)

18 Students conducted a firsthand investigation into the microbial testing of food.

The report's aim is written as:

Aim: *to culture microbes growing on old bread using two different types of medium and to observe their shape, size, colour and surfaces*

How many independent variables were in the firsthand investigation?

(A) 1

(B) 2

(C) 4

(D) 5

(Sample question)

Refer to the following information to answer Questions 19 and 20.

> The intestinal tract of a human foetus is sterile.
>
> After birth, microflora from the mother are transferred to the baby's mouth through close contact. After a year, the microflora of the baby is similar to the mother's, with the baby's immune system ignoring these microbes.
>
> Also during the first year of life, breast milk from the mother provides antibodies to the baby for any disease the mother has already experienced. When breastfeeding ceases, these antibody levels in the baby start to fall.
>
> After the first year, any new species of invading bacteria is treated as a pathogen by the baby's immune system.

19 A medical consequence for six-month-old babies that have only been bottle-fed with formula milk and not breastfed is that

(A) they will not develop microflora.

(B) their immune system will be damaged.

(C) their consumption of milk cannot be quantified.

(D) they will be at increased risk of infectious disease.

(2015 HSC)

20 Strict hygiene practices are followed in the care of newborns, whereas hygiene practices in the care of older babies are less emphasised.

Which of the following is the best reason for this difference?

(A) Vaccinations render personal hygiene unnecessary for older babies.

(B) Procaryotic cells are not identified as antigens in early development.

(C) Antibiotic treatments kill bacterial populations in the digestive system.

(D) Early exposure to pathogens helps to build a strong immune system.

(2015 HSC)

Biology

Section II
Answer Booklet

80 marks
Attempt Questions 21–36
Allow about 2 hours and 25 minutes for this section

Instructions

- Answer the questions in the spaces provided. These spaces provide guidance for the expected length of response.
- Show all relevant working in questions involving calculations.

Please turn over

Question 21 (2 marks) **Marks**

Explain ONE control measure for an infectious disease you have studied. **2**

(Sample question)

Question 22 (2 marks)

Explain why polypeptide synthesis is an important process. **2**

(Sample question)

Question 23 (3 marks)

The diagram shows a model involving DNA.

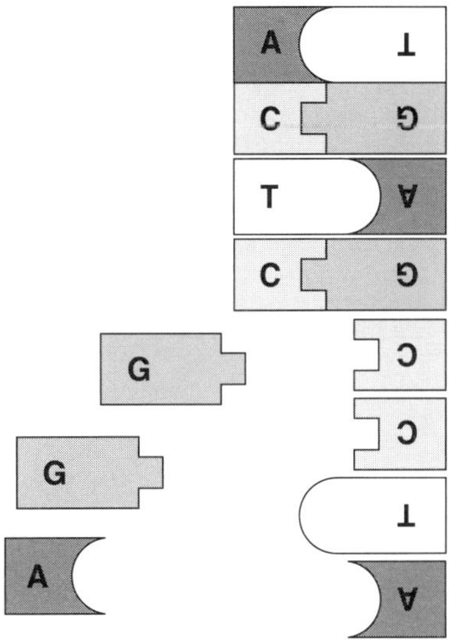

(a) What process is being modelled? **1**

(b) Identify TWO structural features of the DNA molecule which are NOT shown in this model. **2**

(2015 HSC)

Question 24 (7 marks) **Marks**

Your study evaluated the effectiveness of technologies to manage and assist with the effects of disorders in humans.

(a) Name one technology that manages and assists with the effects of disorders in humans and the disorder that it is designed to help. 2

....

....

(b) Outline TWO limitations of the technology. 2

....

....

....

....

(c) Outline TWO advantages of the technology and ONE current advance in the technology. 3

....

....

....

....

(Sample question)

Question 25 (5 marks) **Marks**

A group of students wanted to test whether water purifying tablets were effective in making creek water free from bacteria.

They conducted an experiment using a water sample collected from the creek and found that the tablets were effective.

(a) Describe a means of addressing ONE identified hazard relevant to this investigation. **2**

....

....

....

....

(b) Illustrate the results of this experiment in diagrammatic form. **3**

Use labels to clearly identify the data collected.

(2015 HSC)

Question 26 (5 marks) **Marks**

In your studies you used models to represent and describe processes and concepts.

(a) Explain why models are important in biology. **2**

....

....

....

(b) For three different areas that you have studied, list how you used models and modelling. **3**

....

....

....

(Sample question)

Question 27 (5 marks) **Marks**

(a) Outline TWO methods used to control an epidemic or pandemic. **2**

....

....

....

....

(Sample question)

(b) The steps below show the preparation and use of blood products in the treatment of Ebola Virus Disease. This disease is characterised by significant blood loss.

World Health Organisation Protocol

1. A patient recovers from Ebola Virus Disease.
2. The same patient is disease-free for 28 days.
3. Blood is taken from the patient and screened for transmissible diseases.
4. Plasma is separated from the whole blood.
5. Plasma is transfused into another person with early signs of Ebola Virus Disease.

Explain why this protocol produces an effective treatment for Ebola Virus Disease. **3**

....

....

....

....

....

....

(2015 HSC)

Question 28 (5 marks) **Marks**

A group of students hypothesised that the height of plants decreases with increased elevation.

The students planted ten plant cuttings from the same plant at each of five locations. The locations were at varying elevations in the same mountain range. All the cuttings were provided with the same volume of water on planting, and no fertiliser was applied. The students returned after the same growth period and measured the height of the plants.

The cross-section shown indicates the average height of the plants in metres after the growth period at each location in the mountain range.

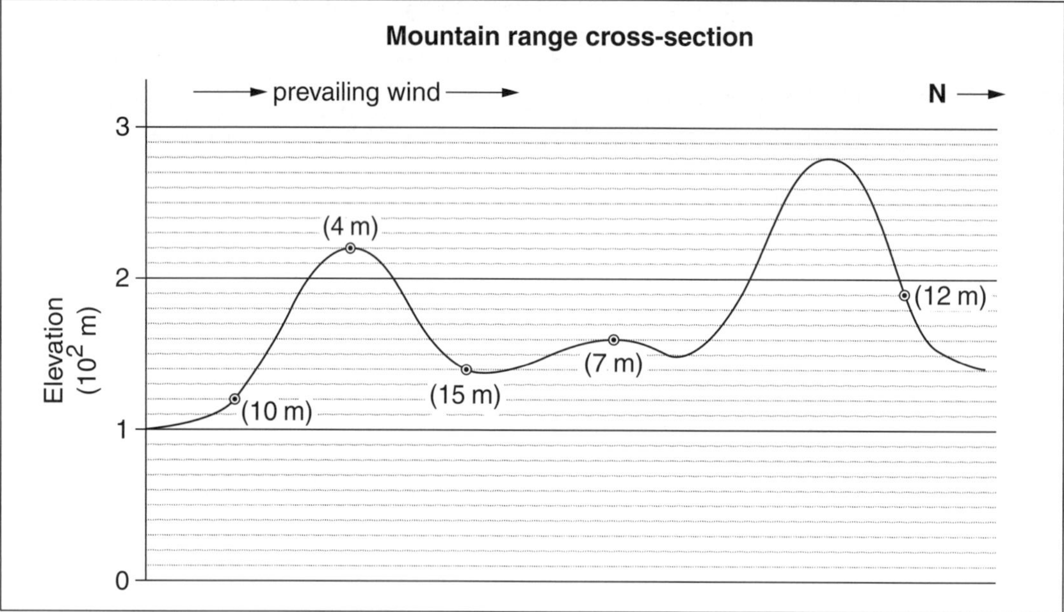

(a) Evaluate the validity of the experiment. **3**

....

....

....

....

....

....

(2015 HSC)

Question 28 continues

Question 28 (continued) **Marks**

(b) Complete both columns of the table to best present the data for the analysis of any trend. **2**

(2015 HSC)

End of Question 28

Question 29 (6 marks) **Marks**

'The application of modern reproductive techniques in plant and animal breeding limits genetic diversity.' **6**

Discuss this statement.

(2015 HSC)

Question 30 (7 marks) **Marks**

The table below simulates data relating to health problems in Australia that caused the highest numbers of disabilities in 2005 and 2016.

Communicable diseases appear in grey text.

Non-communicable diseases appear in italic text.

Data relating to health problems in Australia that caused the highest numbers of disabilities in 2005 and 2016

Ranking in top 10	2005 ranking of diseases	2016 ranking of diseases	Percentage change between 2005 and 2016
1	*Lower back and neck pain*	*Lower back and neck pain*	27.0
2	*Depressive disorders*	*Skin diseases*	19.3
3	*Skin diseases*	*Depressive disorders*	13.4
4	*Migraine*	*Migraine*	17.5
5	*Other musculoskeletal*	*Sense organ diseases*	28.2
6	*Sense organ diseases*	*Other musculoskeletal*	23.4
7	*Anxiety disorders*	*Anxiety disorders*	15.1
8	*Asthma*	*Asthma*	6.9
9	*Drug use disorders*	*Drug use disorders*	26.6
10	*Oral disorders*	*Oral disorders*	38.3

(a) How many communicable diseases are shown in the data? **1**

....

(b) What do you understand by the term 'non-communicable diseases'? **2**

....

....

(c) With reference to the table, describe the per cent changes and rankings from 2005 to 2016 for health problems in Australia that cause the highest numbers of disability. **4**

....

....

....

(Sample question)

Question 31 (8 marks) **Marks**

'Renal dialysis and kidney transplants are very different treatments for the same medical condition. Each treatment was developed from a new application of biological knowledge.' **8**

Justify these statements.

(2015 HSC)

Question 32 (4 marks) **Marks**

Describe TWO technologies that are used to determine inheritance patterns in populations. **4** **4**

(Sample question)

Question 33 (4 marks)

Compare and contrast the nervous pathways with the hormonal system in vertebrates. **4**

(Sample question)

Question 34 (4 marks) **Marks**

The diagram shows two cells taken from the same organism.

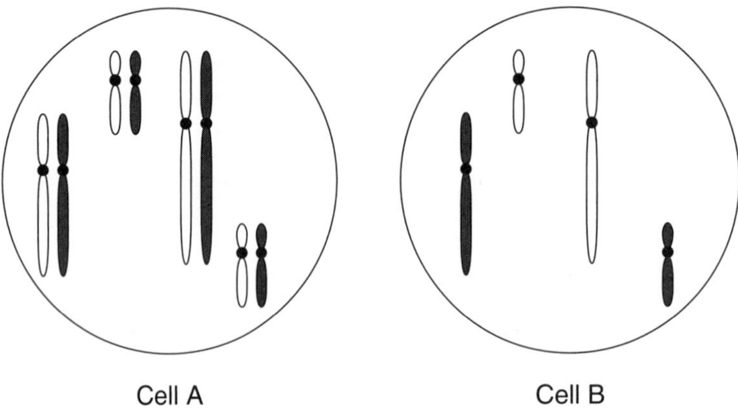

(a) What type of cell is Cell B? **1**

....

(b) Explain the potential effects on this species of a disadvantageous mutation in each of these cell types. **3**

....

....

....

....

(Sample question)

Question 35 (5 marks)

Explain the difference between mutations in 'coding' versus 'non-coding' DNA segments. **5**

....

....

....

....

....

....

(Sample question)

Question 36 (8 marks) **Marks**

Evaluate the impact of scientific knowledge on the manipulation of plant reproduction in agriculture. **8**

(Sample question)

Sample HSC Examination Paper with 2015 HSC Questions

Sample Answers

Section I

1 A All plants and animals have evolved innate immune responses to prevent the entry of and respond to pathogens. Only vertebrate animals have evolved an adaptive immune response to respond to and prevent the entry of pathogens.

2 A This is the only term related to variation in a gene. The other terms relate to chromosome structure.

3 C Sunken stomata prevent the loss of water.

4 D Continuity of a species occurs through effective asexual or sexual reproduction. Both types of reproduction require the replication of genetic material (DNA) and its transfer to the next generation. This transfer of genetic material also ensures the continuity of species. Diversity also plays a role but (A) is not a complete answer.

5 D The type of arrangement is duplication.

6 B Infectious diseases are disorders caused by organisms such as bacteria, fungi, viruses or parasites. Contagious or communicable diseases are those infectious diseases that spread.

7 B The brain sends messages to the effectors in order to promote a response to a stimulus.

8 B If body temperature is low, the hairs stand on end and blood vessels constrict to prevent heat loss.

9 C The size and structure of the pathogen correlate to a fungus.

10 B Many eukaryotic animals and some plants have somatic cells, which are the normal body cells containing chromosomes in pairs, with one member of each pair coming from each parent.

11 D Radiation is a common cause of mutation in genes.

12 B Fewer flies in subsequent generations will eventually reduce or eliminate the population altogether.

13 D By not taking affected fruit carrying the flies to other states and eliminating the new populations, the problem may be contained to local areas only.

14 D This option contains the correct possible combination for tyrosine.

15 C The F2 parents were both heterozygous, so the original parents were both homozygous.

16 D Koch's postulates will only assist in identifying infectious disease.

17 B The helper T cell is releasing cytokine to assist in activation of the macrophage.

18 B Two. The independent variable is the factor that is changed. In this investigation it is the type of medium used.

19 D The babies will have a vulnerable immune system without the benefit of the immunity that is conferred through breast milk from the mother.

20 B The immune systems of newborns is immature and will not work as effectively as in older babies.

Section II

Question 21 (Total 2 marks)

An example of a control measure for malaria is spraying breeding grounds with pesticides. This reduces the number of vectors, thereby decreasing the likelihood of disease spread. *(2 marks)*

Question 22 (Total 2 marks)

Polypeptide synthesis results in the production of components necessary for protein formation. Proteins are vital to the structure and functioning of cells. Inheritance and continuity of a species depends on polypeptide synthesis. *(2 marks)*

Question 23 (Total 3 marks)

(a) The model represents DNA replication and shows unzipping of a double helix, with complementary base pairs moving apart. *(1 mark)*

(b) The model does not represent the sugar-phosphate backbone of the nucleotide attached to each base, nor does it show the helical structure of DNA. The model also does not show the hydrogen bonds that connect the base pairs. *(2 marks)*

Question 24 (Total 7 marks)

(a) Hearing aids are designed to manage and assist with hearing loss. *(2 marks)*

(b) Some limitations of hearing aids are: they do not assist people who have damage to the inner ear or auditory nerve; and because they must provide enough amplification without making sounds painfully loud, some models do not suit some individuals and may need frequent adjustments for different intensities of sounds. *(2 marks)*

(c) The advantage of hearing aids is that they are relatively cheap and easy to install. Hearing aids provide the wearer with better access to sounds and an improved ability to understand speech.

With advances in technologies, hearing aids have become much smaller and digitised. Digital hearing aids can be programmed for specific listening situations and linked to other technologies. *(3 marks)*

Question 25 (Total 5 marks)

(a) One potential hazard that students could encounter in this investigation is accidentally ingesting the water-borne pathogens. One way to minimise this hazard is to wear gloves when handling the water samples and to thoroughly wash hands after the task. *(2 marks)*

(b)

Effectiveness of water purifying tablets on creek water bacteria			
Trial	Number of tablets	Quantity of bacteria present on culture (agar in petri dish)	
		Water sample 1	Water sample 2
A (Control)	0	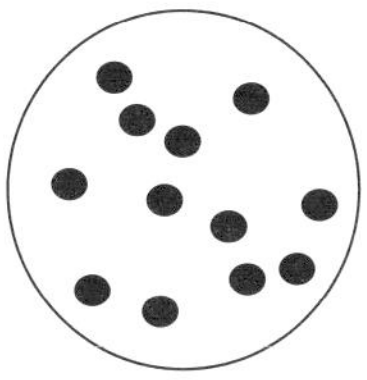	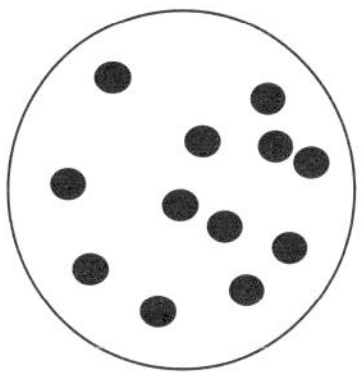
B	1	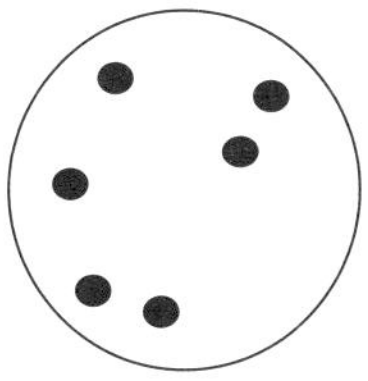	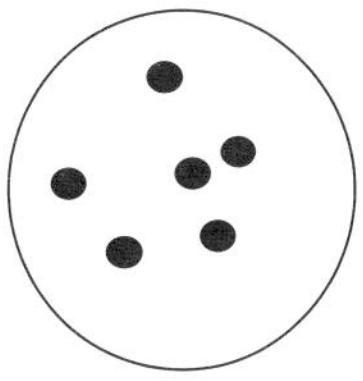
C	2	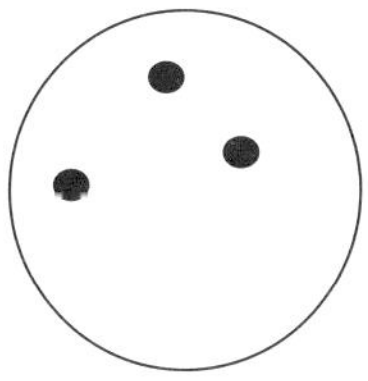	
D	3	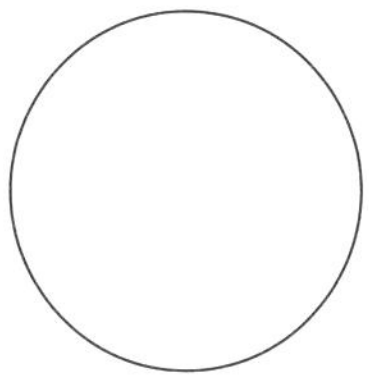	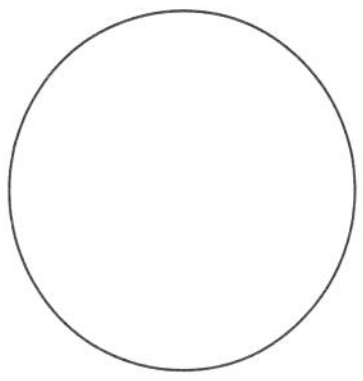

Note to students: diagrams could include agar plates, beakers, flasks, test tubes, tables or graphs. In this example results are represented as a table with diagrams of agar plates. *(3 marks)*

Question 26 (Total 5 marks)

(a) Models can simplify a concept to help in understanding or communication. Models can simplify, clarify or provide an explanation of the workings, structure or relationships within an object, system or idea. *(2 marks)*

(b) Models were used:

- to show the processes involved in cell replication for mitosis and meiosis;
- to show DNA replication using the Watson and Crick DNA model;
- to help explain the innate and adaptive immune systems in humans. *(3 marks)*

Question 27 (Total 5 marks)

(a) Two methods are environmental management, which includes the bio-secure disposal of all contaminated medical materials, and quarantine, which includes the isolation of all suspected individuals. *(2 marks)*

(b) The protocol described represents possible effective treatment of Ebola patients because the globulins transported in plasma can act as antibodies. The original patient with Ebola recovered and was able to produce the antibodies which can then be transferred to the new patient and assist their immune system to fight the disease. *(3 marks)*

Question 28 (Total 5 marks)

(a) While some variables such as volume of water, no fertiliser and taking cuttings from the same plant were controlled, other factors such as exposure to prevailing winds and direction with respect to the sun were not controlled. As both of these factors will affect plant growth, the experiment is not valid. *(3 marks)*

(b)

Elevation (10^2 m)	*Height (m)*
1.2	10
1.4	15
1.6	7
1.9	12
2.2	4

(2 marks)

Question 29 (Total 6 marks)

The application of some modern reproductive techniques in plant and animal breeding can limit the genetic diversity, while other techniques enhance genetic diversity.

If plants or animals are cloned (sheep and pigs) or grafted (fruit trees) then genetic diversity is reduced, as you are manifesting a limited amount of the genes from a species genetic pool. This can have negative consequences if the environment in which the organisms live is a changing one. However, if the sole purpose of breeding organisms in this way is to produce organisms with a limited set of desirable characteristics for agricultural purposes then it can be beneficial to society.

In selective breeding or in-vitro fertilisation (IVF) programs, the diversity is not reduced in the same way because new genetic combinations are being produced and passed, thereby maintaining genetic diversity. Artificial pollination is the transfer of pollen from one plant to the stigma of another plant. Artificial insemination is collecting semen from one animal and transferring it to the reproductive organs of another organism. Examples of where this technology has been used include developing cows with high butter fat in milk or leaner meat, and sheep with better wool quality.

Answer Q29 continues

Answer Q29 (continued)

When organisms are manipulated genetically, such as through the use of recombinant DNA, new genes are introduced into a species' gene pool. The manipulated or inserted genes are passed to the next generation through the transgenic organisms, thereby increasing the genetic diversity within that species' genetic pool. This technology has been used to create frost-resistant strawberries and disease-resistant Bt cotton and corn.

As such, it cannot be said that modern reproduction techniques are limiting genetic diversity; it is quite the opposite effect, as genetic diversity is being created through the use of the reproductive technology. *(6 marks)*

Question 30 (Total 7 marks)

(a) No communicable diseases are shown in the data. *(1 mark)*

(b) Non-communicable diseases are non-infectious diseases caused by something other than a pathogenic organism. *(2 marks)*

(c) The highest percentage of changes between 2005 and 2016 were for oral disorders and sense organ disorders. The rankings changed for four diseases. The skin diseases ranking moved from 3 to 2. The sense organ diseases ranking moved from 6 to 5. The depressive disorders ranking moved from 2 to 3. Other musculoskeletal diseases moved from 5 to 6. *(4 marks)*

Question 31 (Total 8 marks)

Renal dialysis and kidney transplants are used to treat patients with chronic kidney diseases that prevent the purification of their blood. These diseases include glomerulonephritis, polycystic kidneys and diabetes mellitus.

Renal dialysis has been relatively readily available to patients for about the past 50 years. This treatment is based on knowledge of kidney function and osmotic processes in the body. Two distinctly different types of dialysis are possible: haemodialysis, which filters the blood outside the body; and peritoneal dialysis, which operates within the body. Dialysis works on the principles of the diffusion of solutes and ultrafiltration of fluid across a semi-permeable membrane, effectively replacing the work done by the glomerulus in the nephron. Nitrogenous wastes (such as urea) diffuse across a concentration gradient from a high concentration in the blood to a low concentration in the dialysate solution in canister. Today patients can have their own dialysis machines at home and treat themselves with minimal disruption to work and home life.

Answer Q31 continues

Answer Q31 (continued)

With advances in technology and biological understanding of kidney function, it has also become possible to undergo a kidney transplant for the most severe cases. This involves a patient being the recipient of a healthy kidney from a donor that has had the tissue type matched. A foreign tissue contains antigens which stimulate cytotoxic T cells to attack the transplanted organ. Increased understanding of immuno-suppressor drugs, which must be given to the recipient of the transplanted organ, and effective tissue matching techniques has increased the chances of success for transplant patients. The number of people who are willing to donate kidneys has also increased with the knowledge that the procedure will not impact on their lives detrimentally.

Dialysis and transplant options have developed for the same purpose from the increase in biological knowledge over time, with significantly increasing success in each of the treatment options. *(8 marks)*

Question 32 (4 marks)

Two technologies are DNA sequencing and DNA profiling. Both of these technologies depend on a variety of processes that involve the manipulation of DNA. DNA sequencing is used to determine the sequence of nucleotide bases in the genome. This has led to the study of inheritance patterns that has improved our understanding of human evolution. DNA profiling uses technology to determine an individual's DNA characteristics, often for the purposes of paternity testing, genealogy, the identification of remains or in criminal investigations. *(4 marks)*

Question 33 (4 marks)

Both neural and hormonal pathways are internal coordination systems. Receptors and effectors in vertebrates are linked via a control centre by nervous or neural and hormonal pathways. The nervous system provides rapid and short-term coordination of internal organ systems, whereas the hormonal system provides slower and longer-lasting response coordination. *(4 marks)*

Question 34 (4 marks)

(a) Cell B is a gamete. *(1 mark)*

(b) A potential mutation in each of these cell types can have significant effects on a species. If cell A was to have a disadvantageous mutation, the organism might die due to the inability of the affected cell/tissue to function correctly. It could manifest itself as a cancer or possibly organ failure if it is localised. This could mean the affected individual would not be able to reproduce, which could lead to the demise of the species. If cell B was to have a disadvantageous mutation it could be passed on to the offspring. This could result in the organism being sick or unable to look after itself, resulting in decreased numbers of the species. *(3 marks)*

Question 35 (5 marks)
'Coding' sections of DNA segments comprises the genes that are templates for the formation of polypeptides. If a mutation occurs in the coding then the consequences can include point mutations.

'Non-coding' DNA segments in the genome have a range of functions, including regulation of gene expression. The impact of mutations in non-coding segments can range from having no effect to preventing polypeptide synthesis and hence some protein functions. If the non-coding DNA that experiences a mutation is a promoter sequence, DNA polymerase may not be able to bind, thereby preventing the transcription of a polypeptide. *(5 marks)*

Question 36 (8 marks)

Agriculture must become sustainable for the long-term survival of humans and the environment. The ever-growing human population has increased demands on agricultural production. The use of scientific knowledge includes:

- asexual techniques such as cloning plants through plant propagation, which allows the rapid development of plants with known qualities.
- the artificial selection of plants with more desirable seeds and fruit, which has led to a greater diversity of plant foods.
- reproductive technologies used to the manipulate animal reproduction; this includes artificial insemination, in-vitro embryo production and whole organism cloning such as Dolly the sheep.
- gene technology or genetic engineering, which results in genetically modified organisms. An example is the production of Bt cotton that contains a gene from the soil bacterium *Bacillus thuringiensis.* As in the bacterium, the gene in the Bt cotton produces a protein that is toxic to insect larvae. The benefit of this technology is the reduced need for pesticides.

(8 marks)

CHAPTER 6

Sample HSC Examination Paper with 2016 HSC Questions

Biology

General Instructions

- Reading time — 5 minutes
- Working time — 3 hours
- Write using black pen
- Draw diagrams using pencil
- NESA approved calculators may be used
- For questions in Section II, show all relevant working in questions involving calculations

Total marks: 100

Section I – 20 marks

- Attempt Questions 1–20
- Allow about 35 minutes for this section

Section II – 80 marks

- Attempt Questions 21–36
- Allow about 2 hours and 25 minutes for this section

Section I

20 marks

Attempt Questions 1–20

Allow about 35 minutes for this section

---

1 In mammals, hormones are released into the blood by which organ?

(A) Nerves

(B) Brain

(C) Sweat glands

(D) Endocrine glands

(Sample question)

2 Normal microflora are the natural microorganisms that exist on or in the human body. Normal microflora are considered to be pathogenic when they

(A) increase in size.

(B) are resistant to antibiotics.

(C) cause symptoms in a person.

(D) are transferred to another person.

(Sample question)

3 Which cellular structures carry the genetic material DNA?

(A) Daughter cells

(B) Chromosomes

(C) Centromeres

(D) Alleles

(Sample question)

4 What is the name given to the total base composition and arrangement of the DNA in an organism?

(A) Genome

(B) Genotype

(C) Phenotype

(D) Trait

(Sample question)

5 The following diagram shows a visual disorder. What is the name given to his type of visual disorder?

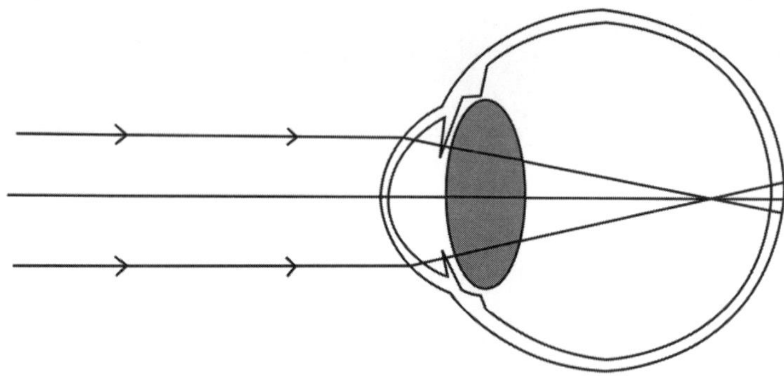

(A) Myopia

(B) Hyperopia

(C) Pretiopia

(D) Fovea

(Sample question)

6 The following diagram provides a simplified model that shows the development of lymphocytes that play a key role in the immune response in the human body.

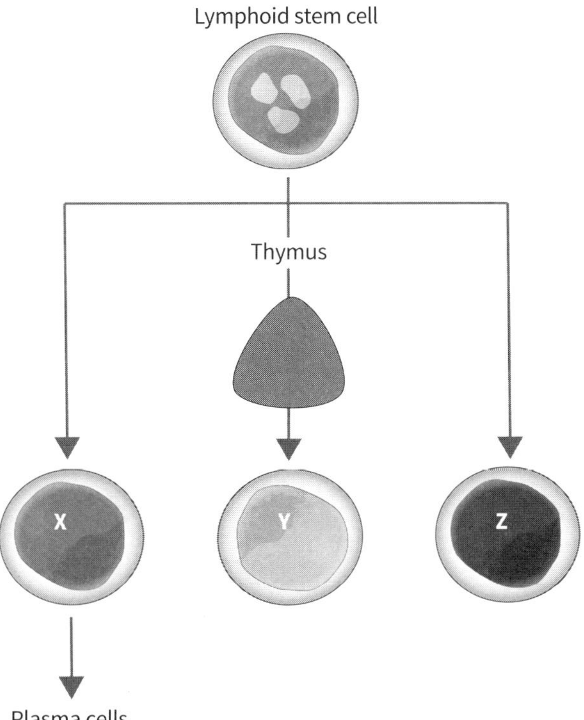

Which of the following shows the correct labelling of the diagram?

	X	*Y*	*Z*
(A)	Natural Killer cells	B cells	T cells
(B)	T cells	Natural Killer cells	B cells
(C)	B cells	T cells	Natural Killer cells
(D)	Stem cells	B cells	T cells

(Sample question)

7 The diagram shows a simplified model of a mammalian nephron and three processes labelled *1*, *2* and *3*.

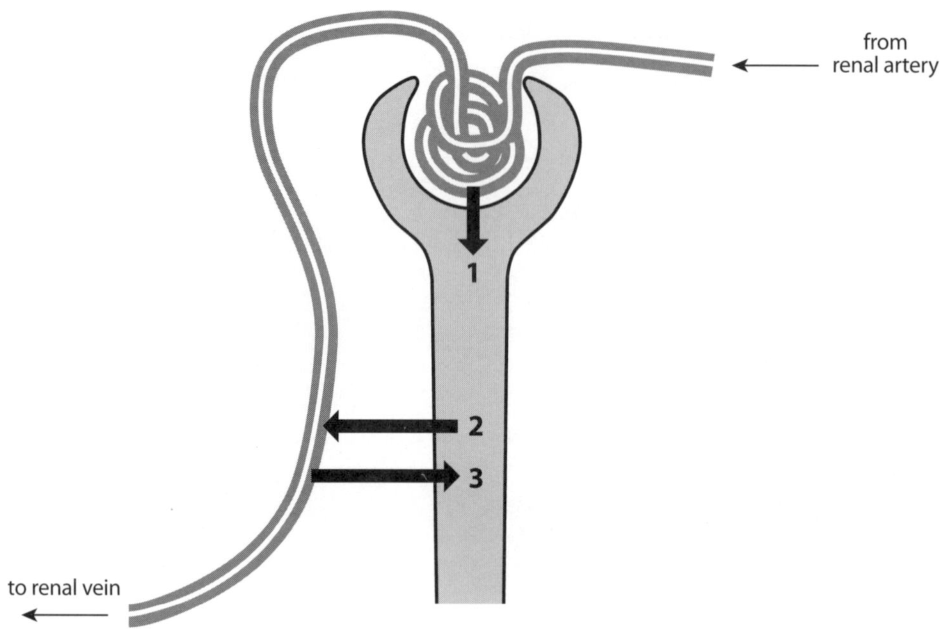

(Source: Adapted from User:Madhero88. March 2010. Physiology of nephron. Available online at https://commons.wikimedia.org/wiki/File:Physiology_of_Nephron.png. Wikimedia Creative Commons 3.0 Unreported CC-BY-3.0)

Which row of the table correctly identifies each of the processes?

	1	*2*	*3*
(A)	Reabsorption	Filtration	Secretion
(B)	Filtration	Reabsorption	Secretion
(C)	Secretion	Reabsorption	Filtration
(D)	Filtration	Secretion	Reabsorption

(2016 HSC)

8 The pedigree shows the inheritance of a characteristic.

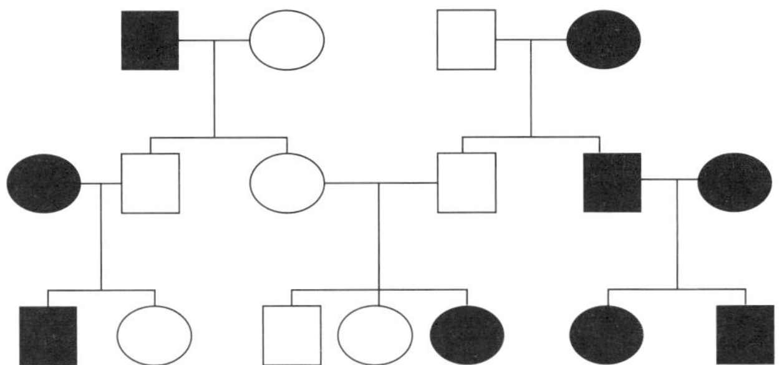

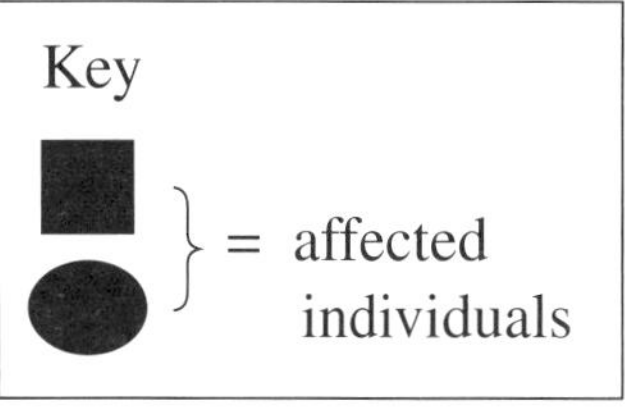

What pattern of inheritance is shown?

(A) Dominant and sex-linked

(B) Recessive and sex-linked

(C) Dominant and not sex-linked

(D) Recessive and not sex-linked

(2016 HSC)

9 Why is passive transport alone inadequate for the production of urine that is high in nitrogenous wastes?

(A) Osmosis cannot target specific solutes.

(B) Solutes cannot move against a concentration gradient.

(C) Solute transport increases at low concentration gradients.

(D) Osmosis moves water from a low concentration to a high concentration.

(2016 HSC)

10 A model of DNA is shown.

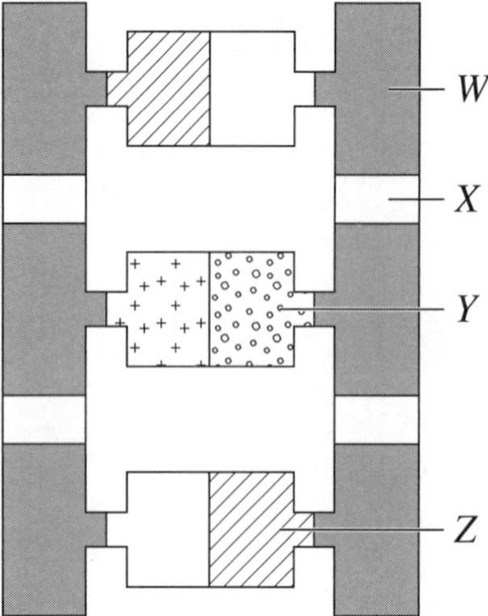

Which row of the table correctly identifies the different components of the model?

	W	*X*	*Y*	*Z*
(A)	Sugar	Phosphate	Adenine	Guanine
(B)	Phosphate	Sugar	Guanine	Cytosine
(C)	Sugar	Phosphate	Adenine	Thymine
(D)	Phosphate	Sugar	Guanine	Thymine

(2016 HSC)

11 Antibodies are molecules released by

(A) memory B cells.

(B) memory T cells.

(C) specialised B cells.

(D) specialised T cells.

(2016 HSC)

12 Some students investigated the response of a plant cutting to an increase in ambient air temperature. They used the apparatus shown.

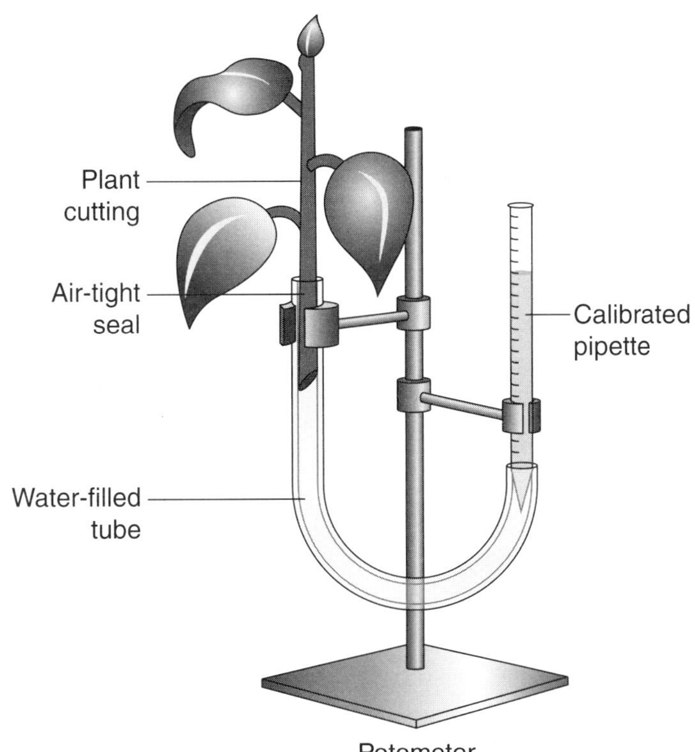

Their data are shown below.

Air temperature (°C)	*Water lost in 60 minutes* (mL)
10	1
18	3
21	4
25	5
30	6

The benefit for the plant of this response would be a decrease in the

(A) rate of transpiration.

(B) rate of photosynthesis.

(C) temperature of the plant.

(D) amount of water in the plant.

(2016 HSC)

Refer to the following information to answer Questions 13 and 14.

The diagram shows some chromosomes during some stages of meiosis.

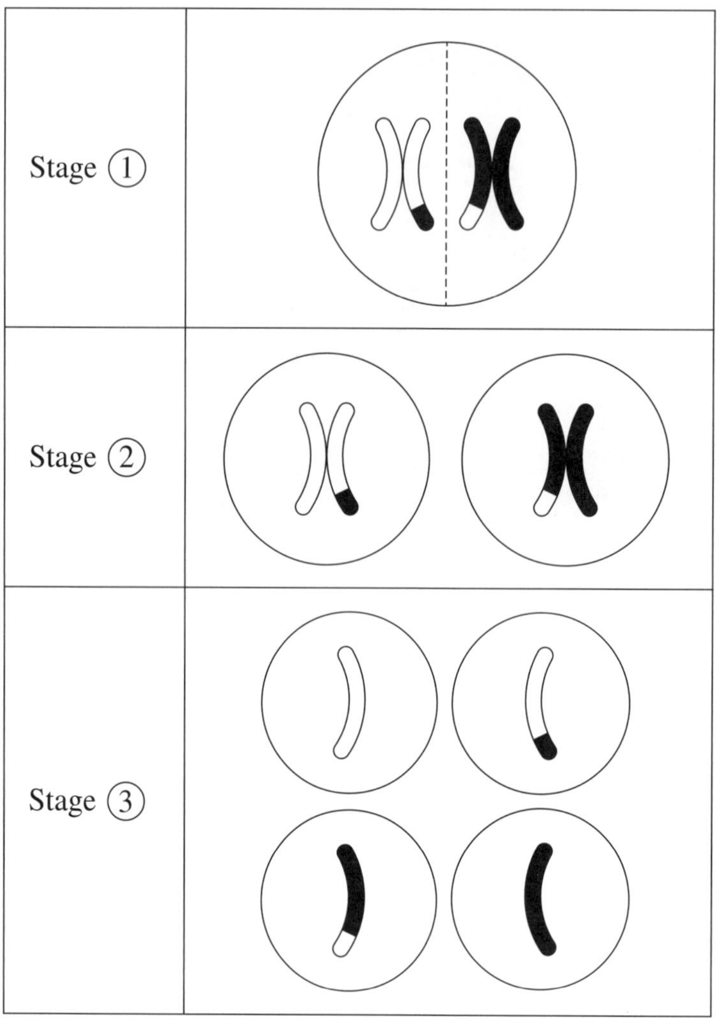

13 When does the segregation of homologous chromosomes occur?

(A) Before stage ①

(B) Between stages ① and ②

(C) Between stages ② and ③

(D) Between stages ① and ② and again between stages ② and ③

(2016 HSC)

14 The chromosomes shown carry

(A) different genes and different alleles.

(B) different genes and the same alleles.

(C) the same genes and different alleles.

(D) the same genes and the same alleles.

(2016 HSC)

15 The table shows four features that can be used to distinguish between types of pathogens.

Which row of the table is correct?

	Feature	*Prion*	*Protozoan*
(A)	Contains DNA	✓	✗
(B)	Cellular	✗	✓
(C)	Can reproduce	✓	✗
(D)	Composed of protein	✗	✓

(2016 HSC)

16 Which of the following is MOST likely to be a response of an Australian plant to a pathogen?

(A) Black marks on the leaves

(B) Development of an abscission layer

(C) Iron deficiency in the soil

(D) Secretion of histamines

(Sample question)

17 Nipah virus (NiV) encephalitis is an emerging infectious disease of public health importance in the South-East Asia region (WHO, 2012). Bangladesh and India have reported human cases of Nipah virus encephalitis. Indonesia, Thailand and Timor-Leste have identified antibodies against NiV in the bat population and the source of the virus has been isolated.

What is the term used to refer to infectious diseases that are transmitted from animals such as bats to humans?

(A) Mobility diseases

(B) Zoonotic diseases

(C) Chronic diseases

(D) Non-communicable diseases

(Sample question)

18 How does the production of a new transgenic species have the potential to alter the path of evolution?

(A) The creation of new genes increases biodiversity.

(B) The removal of genes from a species decreases biodiversity.

(C) The transfer of genes within a species increases biodiversity.

(D) The transfer of genes between two species increases biodiversity.

(2016 HSC)

Refer to the following information to answer Questions 19 and 20.

> Melanomas are characterised by uncontrolled cell division caused by mutations that continue to occur once the tumour has developed. Scientists have discovered that vaccines produced using antigens extracted from the patient's own melanoma cells can be useful in treating melanoma. When injected, the vaccines stimulate an immune response.

19 What can be inferred from the scientists' discovery?

(A) Cancer cells carry unique antigens.

(B) Self-antigens are not present on cancer cells.

(C) The melanoma patient has a dysfunctional immune system.

(D) The body cannot mount an immune response against cancer cells.

(2016 HSC)

20 The effect of the melanoma vaccine is to stimulate

(A) T cells which produce antibodies.

(B) cytotoxic T cells which activate B cells.

(C) cell division to produce more lymphocytes.

(D) production of B cells which destroy melanoma cells.

(2016 HSC)

Biology

Section II
Answer Booklet

80 marks
Attempt Questions 21–36
Allow about 2 hours and 25 minutes for this section

Instructions

- Answer the questions in the spaces provided. These spaces provide guidance for the expected length of response.
- Show all relevant working in questions involving calculations.

Please turn over

Question 21 (2 marks) **Marks**

Explain why, in plants, each plant cell is largely responsible for its own defence response. 2

....

....

(Sample question)

Question 22 (2 marks)

Name an Australian endotherm and describe its response to a decrease in ambient temperature. 2

....

....

....

(Sample question)

Question 23 (5 marks)

(a) What is polymorphism? 1

....

....

(b) What are the most common polymorphisms among humans? 2

....

....

(c) Describe TWO ways that data about polymorphisms is used. 2

....

....

....

(Sample question)

Question 24 (3 marks) Marks

Name an infectious disease and explain how ONE host response is a defence adaptation. 3

....

....

....

....

....

....

(2016 HSC)

Question 25 (6 marks)

Rabies is a disease caused by a virus that affects mammals.

In 1880 Louis Pasteur investigated dogs that were suffering from rabies in order to find the cause. He believed rabies was caused by a microorganism but could not culture it in broth nor observe it under the light microscope. However, he could cause the disease in healthy dogs by injecting them with saliva from infected dogs. He was able to repeat the disease cycle in this way.

(a) Why was Pasteur NOT able to observe the rabies virus? 2

....

....

(b) Explain why Pasteur needed to identify and culture the microorganism in order to meet the scientific standards for establishing the cause of rabies. 4

....

....

....

....

....

....

....

....

(2016 HSC)

Question 26 (7 marks) **Marks**

Students conducted preliminary experiments across different species to analyse their DNA base composition.

The table shows the experimental data collected.

Species	*% Adenine*	*% Guanine*
A	38	12
B	26	22
C	8	40
D	20	32
E	33	18

(a) On the grid below, plot the % Adenine vs % Guanine of the species analysed AND draw a suitable line of best fit. **3**

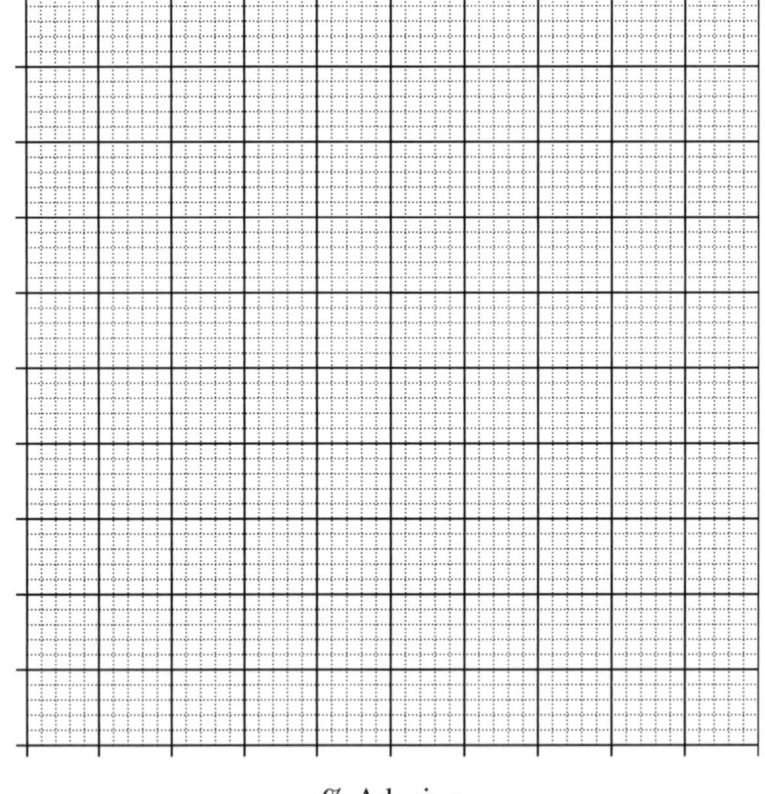

Question 26 continues

Question 26 (continued) **Marks**

(b) Identify the relationship shown by the data. **1**

....

....

(c) Explain the relationship shown by the data. **3**

....

....

....

....

....

....

End of Question 26

(2016 HSC)

Question 27 (4 marks)

Your study collected secondary-sourced data and information relating to a number of topics.

(a) Describe how you collected secondary-sourced data and information. Give ONE example from one area of study. **2**

....

....

....

....

(b) How did you assess that both the data and information were relevant, valid and reliable? **2**

....

....

....

....

(Sample question)

Question 28 (6 marks) **Marks**

Bacteriophages are viruses that use bacteria as host cells. One type of bacteriophage infects only one species of bacterial cell. The infected cells are destroyed as the bacteriophage particles burst out of them, completing their life cycle.

Bacteriophages can be used to fight bacterial infections. As bacteria evolve, bacteriophages also evolve.

(a) Explain TWO advantages of using bacteriophage treatment compared to antibiotic treatment for bacterial infections. **4**

....

....

....

....

....

....

....

....

....

....

(b) Describe a possible disadvantage of using bacteriophage treatment. **2**

....

....

....

....

(2016 HSC)

Question 29 (7 marks) **Marks**

(a) What is the difference between asexual and sexual reproduction? **3**

....

....

....

(b) Construct a table to summarise the advantages and disadvantages of asexual and sexual reproduction. **4**

(Sample question)

Question 30 (5 marks) **Marks**

Explain why the combined use of quarantine and vaccination programs is a more effective way of controlling disease than using only one of these strategies. **5**

(2016 HSC)

Question 31 (8 marks)

(a) Explain the difference between intracellular and extracellular pathogens and provide an example of each. **4**

(b) Explain the TWO different adaptive immune responses in the human body in response to intracellular and extracellular pathogens. **4**

(Sample question)

Question 32 (3 marks) **Marks**

The diagram shows a step in protein synthesis.

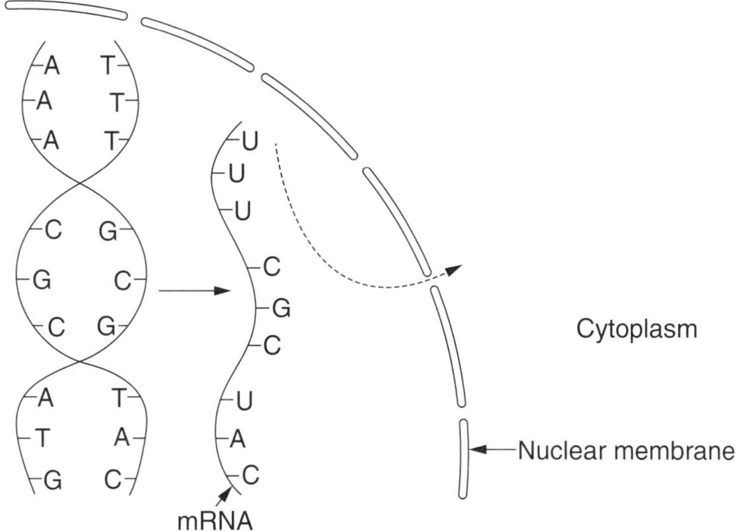

(a) Name the process by which mRNA is produced. 1

....

(b) Explain why the mRNA moves in the direction shown. 2

....

....

(Sample question)

Question 33 (4 marks)

(a) What is 'population genetics'? 1

....

....

(b) Briefly outline three ways data and studies from population genetics are used. 3

....

....

....

(Sample question)

Question 34 (4 marks) **Marks**

Use examples to explain the possible effect on biodiversity of using biotechnology in agriculture. **4**

(Sample question)

Question 35 (7 marks)

Extensive studies of how humans detect both visual and auditory stimuli led to models that shaped our understanding of vision and hearing. These models have been used to design and develop technologies to assist people who experience hearing or visual disorders.

Justify these statements. **7**

(Sample question)

Question 36 (7 marks) **Marks**

'Genes influence proteins and proteins influence genes.'

Evaluate this statement with reference to the structure and function of genes and proteins. 7

(Sample question)

Sample HSC Examination Paper with 2016 HSC Questions

Sample Answers

Section I

1 D In mammals, hormones are released into the blood by endocrine glands.

2 C Microorganisms can become pathogenic in a number of ways, including the presence of large numbers of the pathogen, which means there are too many for the host tissue to function normally. If normal conditions change, microflora can increase in number thereby causing symptoms.

3 B Chromosomes carry the genetic material DNA so that it can be passed on, intact, to daughter cells chromatids, which are two identical copies of a chromosome that are formed at the beginning of cell division and joined together through most of the cell division at a structure called a centromere.

4 A The total base composition and arrangement of the DNA in an organism is called its genome. The genetic composition is called the genotype. The specific observable or measurable characteristics or traits of an organism is called its phenotype.

5 A Myopia is a visual disorder in which the light is focused in front of the retina instead of directly on it.

6 C The presence of an antigen stimulates B cells to make more B cells and specialist cells called plasma cells. T cells complete their development in the thymus. Natural killer cells are large, granular lymphocytes.

7 B Filtration, reabsorption and secretion is the only correct sequence of processes.

8 D Sex-linked traits are located on the 'X' and so are either present or not.

9 B B is the best of the options given. D doesn't make sense as osmosis moves water from a high to low concentration (or from low to high solute) and the other options are incorrect.

10 A Sugars (*W*) attach to bases and the two bases cannot be complementary.

11 C Specialised B cells or plasma cells produce antibodies.

12 C The benefit of increased ambient temperature is a decrease in the temperature of the plant as increased water loss will help regulate the temperature of the plant.

13 B Homologous chromosomes are only in stage 1 so segregation occurs during meiosis 1 (between stages 1 and 2).

14 C Given the chromosomes are homologous, by definition they carry the same genes but there has been an exchange of alleles.

15 B Protozoans are cellular organisms capable of reproduction.

16 B One plant response to a pathogen is the development of an abscission layer around the site of infection that causes the infected tissue to separate from healthy tissue. Black marks are often symptoms of disease. Iron deficiency can be a cause. Histamines are released by special types of white blood cells and therefore are a response by animals.

17 B Zoonotic diseases are diseases that are spread from animals to humans.

18 D Transgenesis relates to the transfer of genes between species.

19 B If self-antigens are not present in the cancer cells then the vaccine can create an immune response against them.

20 C The vaccine would promote development of lymphocytes.

Section II

Question 21 (Total 2 marks)

Unlike animals, plant immune systems lack mobile cells and a circulatory immune system and therefore each plant cell is largely responsible for its own defences against disease. *(2 marks)*

Question 22 (Total 2 marks)

A bilby. (Endotherms are animals that can generate their own body heat internally.) A response to a decrease in ambient temperatures would be a response to keep warm and not lose heat from the body; for example, curling up in a ball or huddling together to decrease surface temperature. *(2 marks)*

Question 23 (Total 5 marks)

(a) 'Polymorphism' refers to genes in a population that occur with multiple variants or alleles. *(1 mark)*

(b) Single nucleotide polymorphisms are the most common genetic variation among humans. Single nucleotide polymorphisms are variations in one nucleotide base, such as thymine being substituted by cytosine on a particular point in the DNA. *(2 marks)*

(c) Data analysis of information about allele frequency, variations and changes contributes to decisions about conserving endangered species and provides evidence for the genetic component cause of some diseases. *(2 marks)*

Question 24 (Total 3 marks)

Responses can vary but must name an infectious disease.

An infectious disease caused by *Vibrio cholerae* is cholera. The host response is to produce antibodies to the toxin. If the patient survives, future exposure to the same strain results in resistance. This is an example of antibody-mediated immunity and results in increased resistance which is of adaptive advantage as it can result in the survival of the individual.

(*3 marks*)

Question 25 (Total 6 marks)

(a) The rabies virus is far too small to be seen with the naked eye or even a simple light microscope. The technology available to Pasteur was not as advanced as it is today; he needed an electron microscope. *(2 marks)*

(b) Pasteur needed to culture the microorganism so that he could identify it and link it directly to the cause of symptoms in individuals. The scientific standard for identifying a disease is based on Koch's postulates:

- the disease must be present in every host with the disease
- the organism needs to be isolated from the host and made into a pure culture
- a healthy host without symptoms must develop the same symptoms after inoculation with the pure culture
- the microorganism needs to be isolated from the new host and a second pure culture made that is the same as the original culture.

So by culturing the rabies virus in unaffected dogs Pasteur was able to determine the cause and produce a vaccine to prevent others from developing the disease if exposed to the same virus. *(4 marks)*

Question 26 (Total 7 marks)

(a)

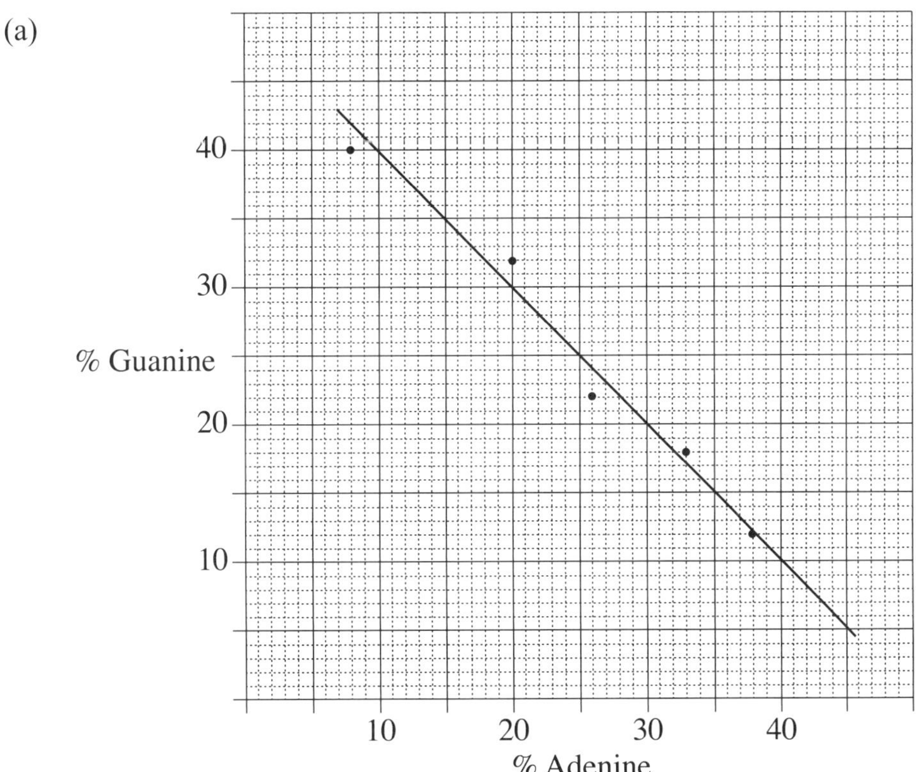

(3 marks)

(b) As the percentage of adenine increases, the percentage of guanine decreases. *(1 mark)*

(c) As the percentage of adenine increases, you would expect the percentage of thymine to increase as these are a complementary base pair. Guanine and cytosine are complementary, not adenine and guanine. So the graph indicates that some species have a higher adenine/thymine proportion of base pairs than cytosine/guanine. *(3 marks)*

Question 27 (Total 4 marks)

(a) Secondary data and information are sourced from factsheets, journal articles, the worldwide web (Internet), books, research articles, and government and non-government agencies. For example, data and facts relating to infectious and non-infectious diseases can be sourced from state and national departments of health and from the World Health Organisation. *(2 marks)*

(b) For assessing validity and reliability look for references, currency (date of publishing), what is the source? Is it a credible source such as a government organisation? If it is from a journal article, look at who is writing the article and what is its purpose. For assessing relevance, consider whether the information answers the inquiry question. Does it address the content or topic area and is it supported by information from other sources? *(2 marks)*

Question 28 (Total 6 marks)

(a) An advantage of using bacteriophage treatment is that antibiotic resistance could be avoided, as no antibiotics are used in the treatment of the disease, thereby preventing overuse or misuse of antibiotics. A second advantage of using bacteriophage treatment is that each bacteriophage is specific to a particular bacteria. If introduced into an infected patient the bacteriophage can destroy the pathogenic bacteria as they use them to reproduce. Once the pathogenic bacteria are destroyed the bacteriophage would also reduce in number as they would no longer have a suitable host to complete their life cycle. *(4 marks)*

(b) A possible disadvantage of using bacteriophage treatment is the unknown effect the residual bacteriophage may have on the body until it completely disappears and the possibility that it may evolve to use other host cells other than the original bacteria it was used to destroy. *(2 marks)*

Question 29 (Total 7 marks)

(a) Asexual reproduction involves a single parent producing genetically identical offspring. Sexual reproduction involves parents of different genders producing haploid gametes that unite to form the diploid zygote. *(3 marks)*

(b) The table summarises the advantages and disadvantages of asexual and sexual reproduction.

	Asexual reproduction	*Sexual reproduction*
Advantages	• Only one parent is needed • The rapid population growth of identical members	• Generally requires two parents of different sex to produce viable gametes • Requires fertilisation and therefore specialised structures and/or water
Disadvantages	• The lack of genetic diversity (biodiversity) within population	• Results in genetic diversity (biodiversity) in a population

(4 marks)

Question 30 (Total 5 marks)

The combined use of quarantine and vaccination programs is a more effective way of controlling disease than using either exclusively because the herd immunity resulting from effective vaccination programs can be backed up by isolation of discrete occurrences of the disease. An effective vaccination program can ensure that individuals that are weak, such as the very young with under-developed immune systems that cannot be immunised or the frail and elderly with weakened immune systems, can be protected by the herd immunity. This means that there is enough of a barrier in the population or community to prevent the disease from being passed on should an infected individual be introduced. If infected individuals are identified, they should be immediately quarantined to prevent them mixing with the community/population so that they do not come into contact with the vulnerable individuals and the disease then is less likely to be spread. A good recent example of this was the isolation of Ebola victims with the simultaneous research into possible vaccinations to limit the spread of the extremely infectious disease. (*5 marks*)

Question 31 (Total 8 marks)

(a) Intracellular pathogens grow and reproduce inside the host cell. All viruses are intracellular pathogens. Extracellular pathogens grow within the tissues and tissue fluids in the host body and do not need to be in the host cell for growth and reproduction. Most bacteria are extracellular pathogens. *(4 marks)*

(b) If the pathogen is outside the cells in the tissues, body fluids or blood, there is a humoral immune response. B cells are involved in the humoral response. If the pathogen is inside the cells, a cell-mediated immune response is needed. This response involves T cells. *(4 marks)*

Question 32 (Total 3 marks)

(a) Transcription (*1 mark*)

(b) mRNA moves across the nuclear membrane out of the nucleus into the cytoplasm as it carries the DNA base template to the ribosomes located within the cytoplasm for translation to occur. (*2 marks*)

Question 33 (Total 4 marks)

(a) Population genetics is the study of allele frequencies in populations and how these change over time. *(1 mark)*

(b) Data and studies of population genetics are used:

- in conservation management
- to determine the inheritance of a disease or disorder and
- in researching human evolution. *(3 marks)*

Question 34 (Total 4 marks)

The use of biotechnology in agriculture has reduced biodiversity on Earth. It has led to changes to ecosystems and reduced the number of species and varieties or breeds of plants and animals as the increased volumes of monocultural, commercial ones allow those to dominate. Examples of reduced biodiversity include the following:

- GM plants containing genes that produce insecticides may reduce biodiversity because of declining populations of both target and non-target insects.
- A reduction in weeds as a result of herbicide-resistant crops has impacted populations of seed-eating birds. Insect-eating birds are potentially at risk from the use of GM crops with inbuilt insecticides. *(4 marks)*

Question 35 (Total 7 marks)

In order to design and develop technologies to assist people who experience disorders, medical scientists, engineers and mathematical modellers must understand the structures and functions of the ears and eyes and the causes and effects of the associated disorders. For example, good hearing requires good conduction of the sound waves through the outer ear, proper mechanical working of the ear ossicles, healthy nerve cells and fluids in the inner ear and functional nerve pathways for the signals to the brain. If anything goes wrong with these structures and their functions, a person will have some kind of hearing loss. Technology designers and developers not only need to understand the processes of sound conduction in the ear but also understand how sound travels and develop prototypes and models that will assist in the hearing loss. The technologies will depend on the type of hearing loss. For example, bone-conduction hearing aids are external devices designed for people with conductive or unilateral hearing loss (hearing loss in only one ear) who get no benefit from wearing conventional hearing aids. Bone-conduction hearing aids create vibrations that move across the skull.

The structures of the eye relate to its function of admitting light, refracting and focusing light to form an image, and converting that image into nerve impulses that are then conveyed to the brain for interpretation. To design technologies to assist people with visual disorders, designers need to have models for how light is reflected and refracted and how the eye functions. The technologies will depend on the type of visual disorder and the eye structure involved. For example, cataract surgery involves the removal of the opaque lens and its replacement with a clear artificial replacement lens called an intraocular lens. *(7 marks)*

Question 36 (Total 7 marks)

Genes are discrete segments of the macromolecule deoxyribonucleic acid (DNA) found on chromosomes and the code for polypeptides. DNA is a double-helical molecule consisting of two strands of sugar and phosphate groups that are joined by pairs of nitrogenous bases (adenine A, thymine T, guanine G and cytosine C). Protein synthesis is the process by which the gene is transcribed and translated within the cell to result in the production of a chain of amino acids that comprise a polypeptide. The specific pairing of the DNA bases (A with T, G with C) in sets of three bases results in triplets that each specifically code for a particular amino acid in the polypeptide sequence. Proteins are macromolecules that are made up of one or more polypeptide chains. Each polypeptide chain becomes folded in a specific manner and there are several polypeptide chains in a particular protein. The actual structure of the protein determines its function. For example, some proteins act as enzymes catalysing specific chemical reactions within the organism. Enzymes bind with their specific substrate at their active sites to promote the reaction with less energy requirements so that the reactions can occur at sufficient speed to support life. If the active site of the enzyme does not have the correct structure it cannot carry out its role. This may occur if there has been a mutation in the DNA of the gene coding for that specific enzyme protein. Thus there is a high level of dependence of proteins, whether they be structural or functional within an organism, on the gene/s that coded for their structure. Genes work in a coordinated manner, especially in embryonic development, where cascades of genes need to operate in sequence to produce limbs and other skeletal and neurological structures. The switching off and on of genes may be a response to particular gene products, proteins. Hormones produced from the activation of genes during puberty may in turn promote changes that result in the activation of other genes. Growth hormone will prompt the action of genes to result in the coordinated changes of a spurt of growth. While it is obvious that proteins depend on genes, there is also some evidence of examples of when genes depend on proteins. (*7 marks*)

CHAPTER 7

Sample HSC Examination Paper with 2017 HSC Questions

Biology

General Instructions

- Reading time — 5 minutes
- Working time — 3 hours
- Write using black pen
- Draw diagrams using pencil
- NESA approved calculators may be used
- For questions in Section II, show all relevant working in questions involving calculations

Total marks: 100

Section I – 20 marks
- Attempt Questions 1–20
- Allow about 35 minutes for this section

Section II – 80 marks
- Attempt Questions 21–36
- Allow about 2 hours and 25 minutes for this section

Section I

20 marks
Attempt Questions 1–20
Allow about 35 minutes for this section

---

1 What is the name of the process that enables organisms to maintain a relatively stable internal environment?

A. Osmosis

B. Adaptation

C. Homeostasis

D. Active transport

(2017 HSC)

2 Which of the following body systems is involved in detecting and responding to environmental changes?

A. Circulatory

B. Digestive

C. Excretory

D. Nervous

(2017 HSC)

3 Which scientist contributed rules of procedure for showing that a particular microorganism is the cause of a particular disease?

A. Robert Koch

B. Louis Pasteur

C. Maurice Wilkins

D. Frank Mcfarlane Burnett

(Sample question)

4 What is the role of lymphocytes in the body?

A. They fight infection.

B. They initiate blood clotting.

C. They transport oxygen around the body.

D. They transport carbon dioxide around the body.

(2017 HSC)

5 The following graph shows the changes in hormone levels in human females during pregnancy.

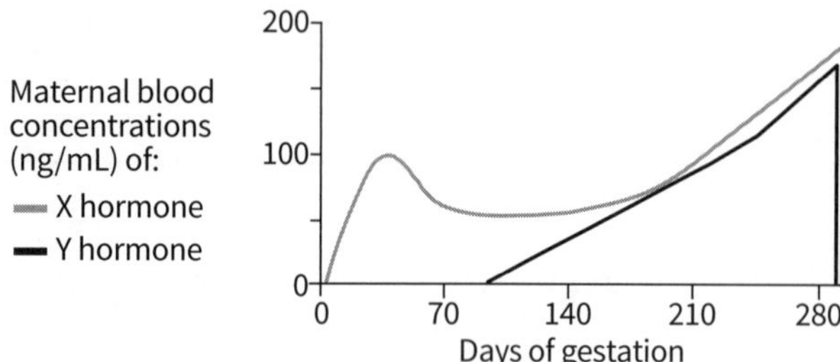

What hormone is represented by the grey line on the graph?

A. Oestrogen

B. Menstrual hormones

C. Progesterone

D. Endocrines

(Sample question)

6 What name is given to the process whereby a white blood cell engulfs a microorganism?

A. Infection

B. Inflammation

C. Phagocytosis

D. Vaccination

(2017 HSC)

7 Which waste product does renal dialysis remove?

A. Urea

B. Urine

C. Lipids

D. Vitamins

(2017 HSC)

8 What is a polypeptide?

A. A peptide bond

B. A long chain of peptide bonds from a code

C. A long chain of amino acids that make up proteins

D. A long chain of proteins

(Sample question)

9 An experiment was planned to investigate the effect of the enzyme, amylase, on starch. The following combination of test tubes was considered.

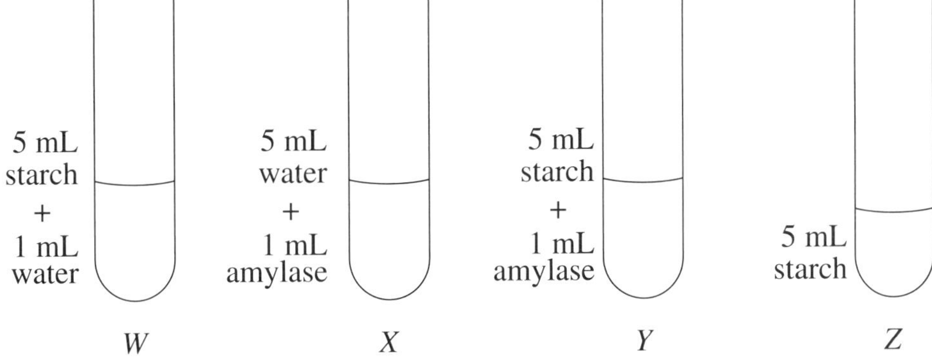

Two drops of iodine will be added to each test tube.

Which combination of test tubes would ensure that the experiment is valid?

	Experiment	*Control*
A.	*Y*	*W*
B.	*Y*	*Z*
C.	*X*	*W*
D.	*X*	*Z*

(2017 HSC)

10 Which statement best describes the role that insulin and glucagon play in the negative feedback loops that maintain homeostasis for glucose?

A. Insulin lowers blood glucose levels by facilitating glucose transport from the cells into the blood. Glucagon raises blood glucose levels by metabolising glycogen into glucose.

B. Insulin lowers blood glucose levels by facilitating glucose transport from the blood into cells. Glucagon raises blood glucose levels by metabolising glycogen into glucose.

C. Insulin raises blood glucose levels by facilitating glucose transport from the blood into cells. Glucagon lowers blood glucose levels by metabolising glycogen into glucose.

D. Both insulin and glucagon lower blood glucose levels by facilitating glucose transport from the blood into cells.

(Sample question)

11 A student was asked to complete a table showing whether T cells and B cells have particular characteristics.

Which row did the student complete correctly?

	Characteristic	*T cell*	*B cell*
A.	Produces plasma cells	✓	✓
B.	Produces antibodies that are released in body fluids	✓	✗
C.	Cell surface receptor can recognise a specific antigen	✓	✓
D.	Forms clones once stimulated	✗	✓

(2017 HSC)

12 What is the probability of producing a tall pea plant when a heterozygous tall pea plant is crossed with a homozygous short pea plant?

A. 0%

B. 50%

C. 75%

D. 100%

(2017 HSC)

13 A section of DNA has the following nucleotide sequence.

AGG TCT CAG ATC

What is the nucleotide sequence of the newly-made strand following DNA replication?

A. AGG TCT CAG ATC

B. AGG UCU CAG AUC

C. UCC AGA GUC UAG

D. TCC AGA GTC TAG

(2017 HSC)

14 Which statement correctly describes fungi and protozoans?

A. Fungi and protozoans are unicellular.

B. Fungi and protozoans have chloroplasts.

C. Fungi have a cell wall and protozoans do not.

D. Fungi are procaryotic and protozoans are eucaryotic.

(2017 HSC)

15 A plant breeder crosses two plants of the same species. They are both pure-breeds for flower colour. The colour of the flowers of all the offspring is the same, but different to that of the parents.

What type of inheritance does this result show?

A. Sex-linked

B. Recessive

C. Dominance

D. Co-dominance

(2017 HSC)

16 What type of mutations can be passed onto offspring and therefore alter the allele frequency in a gene pool?

A. Somatic mutations

B. Mutagens

C. Naturally occurring mutagens

D. Germ-line mutations

(Sample question)

17 In 2015 there were 16 400 deaths due to diabetes in Australia with approximately half of these deaths due to type 2 diabetes. In 2016 diabetes was Australia's seventh leading cause of death, accounting for 3% of all deaths (ABS, 2016).

This data is an example of which of the following?

A. Mortality rates

B. Morbidity rates

C. Co-morbidity rates

D. Incidence rates

(Sample question)

18 Which feature of DNA is important for cell replication?

A. Polyploidy

B. Cytoplasmic streaming

C. DNA replication

D. Mitosis

(Sample question)

19 A zebronkey hybrid is the result of crossing a male zebra which has 44 chromosomes with a female donkey which has 62 chromosomes.

How many chromosomes will the zebronkey have?

A. 53

B. 75

C. 84

D. 106

(2017 HSC)

20 Which of the following is an example of the future use and application of biotechnology?

A. Developing cultures of human stem cells for use in drug trials

B. Fermentation for cheese making

C. Medical applications to produce vaccines

D. Aquaculture for selective breeding

(Sample question)

Biology

Section II
Answer Booklet

80 marks
Attempt Questions 21–36
Allow about 2 hours and 25 minutes for this section

Instructions

- Answer the questions in the spaces provided. These spaces provide guidance for the expected length of response.
- Show all relevant working in questions involving calculations.

Please turn over

Question 21 (4 marks) **Marks**

(a) Give TWO benefits of epidemiological studies. **2**

....

....

....

(b) Provide ONE example of an epidemiological study and a benefit that came from the study. **2**

....

....

....

(Sample question)

Question 22 (4 marks)

For each type of disease in the following table, name a specific disease and its cause. **4**

Type of disease	*Name of disease*	*Cause of disease*
Infectious disease		
Non-infectious disease		

(q22(a), 2017 HSC)

Question 23 (5 marks) **Marks**

The following flowchart is a simplified model that shows innate immune defences and responses of animals to pathogens.

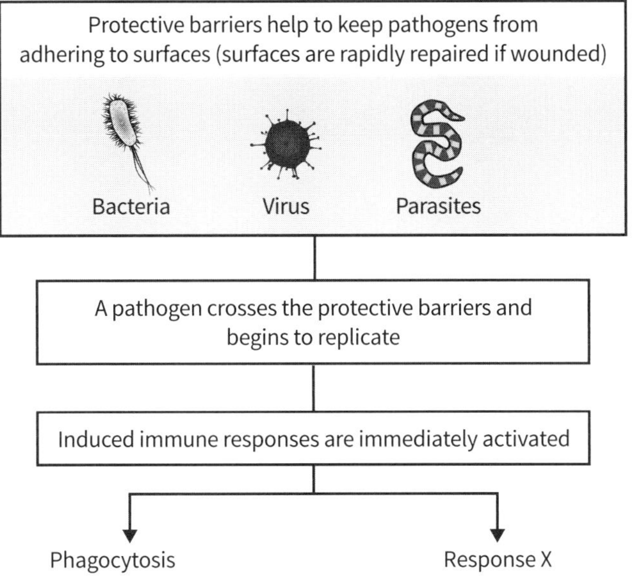

(a) Name the response labelled X. **1**

....

(b) Outline the role of the response labelled X. **2**

....

....

(c) Describe ONE chemical and ONE physical change in the body as a result of the response labelled X. **2**

....

....

....

....

(Sample question)

Question 24 (7 marks) **Marks**

(a) Three genes are arranged along a homologous pair of chromosomes as shown.

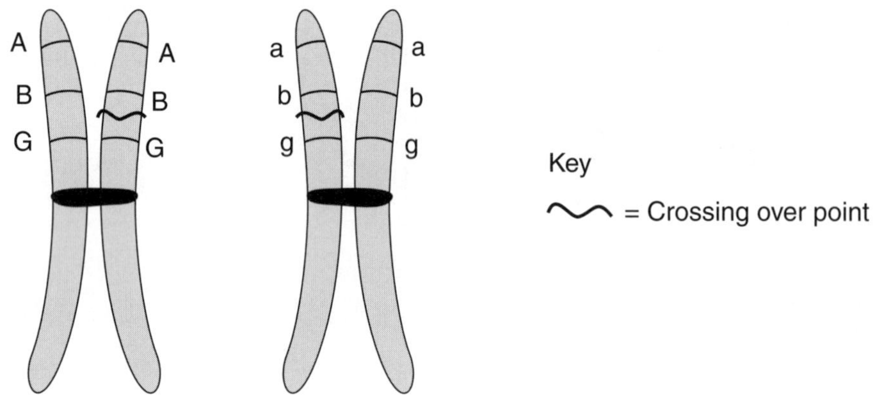

(i) What is the individual's genotype before crossing over occurs? **1**

....

(ii) Label, on the diagram below, the alleles after crossing over has occurred. **1**

(b) Explain the effect of independent assortment of chromosomes on the genotype of the offspring. **2**

....

....

....

....

(2017 HSC)

Question 24 continues

Question 24 (continued) **Marks**

(c) Explain how crossing over produces variation in gametes. **3**

(Sample question)

Question 25 (4 marks)

(a) Explain why the recognition and protection of Indigenous cultural and intellectual property is important. **2**

(b) Describe one process for ensuring the recognition and protection of Indigenous cultural and intellectual property. **2**

(Sample question)

Question 26 (4 marks) **Marks**

A controlled experiment was performed to investigate the effect of substrate concentration on the rate of an enzyme-catalysed reaction. Data were collected and are presented in the graph.

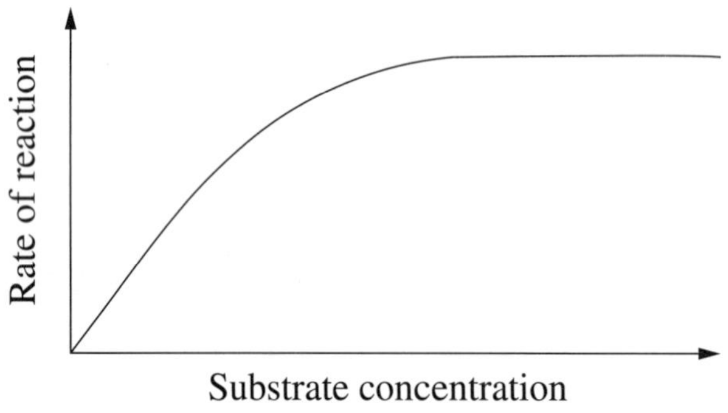

Rate of reaction

Substrate concentration

(a) What is the independent variable in this experiment? **1**

....

(b) Explain the trends shown in the graph. **3**

....

....

....

....

....

....

(2017 HSC)

Question 27 (5 marks) **Marks**

Your study investigated procedures that can be employed to prevent the spread of infectious disease.

(a) What do you understand by the term ‘prevention’? **1**

....

(b) Describe TWO different procedures employed to prevent disease. For each, provide a specific example of a disease that is prevented from spreading. **4**

....

....

....

....

....

....

(Sample question)

Question 28 (3 marks) **Marks**

A pedigree chart of an inherited characteristic is shown. **3**

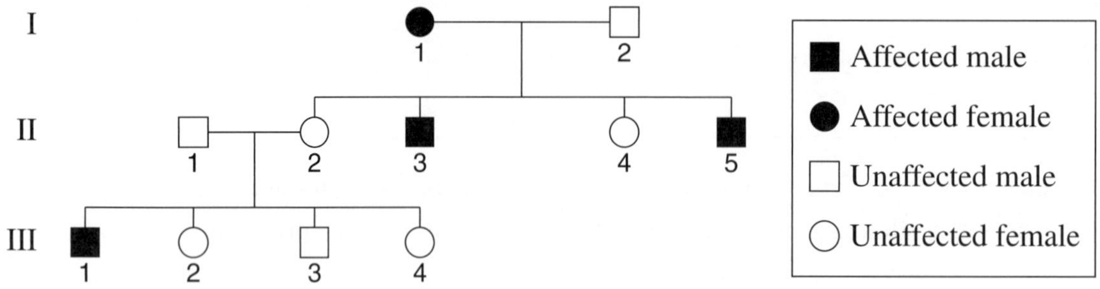

I
1 2
II
1 2 3 4 5
III
1 2 3 4

Affected male
Affected female
Unaffected male
Unaffected female

Subsequent genetic analysis showed I-2 does not have the recessive allele.

Explain the inheritance of this characteristic.

....

....

....

....

....

....

....

....

(2017 HSC)

Question 29 (5 marks) **Marks**

Using an example of a human disease, outline the reason why health systems focus on prevention of disease as much as treatment and management. **5**

....

....

....

....

....

....

....

....

(Sample question)

Question 30 (6 marks)

(a) What is the role of ONE type of T lymphocyte in the immune response? **2**

....

....

....

....

(2017 HSC)

(b) Explain how the immune system responds after primary exposure to a pathogen, including acquired immunity. **4**

....

....

....

....

....

....

....

(Sample question)

Question 31 (8 marks) **Marks**

Advances in technologies are making it increasingly possible to treat and control many human diseases and disorders.

Describe a technology that is used in each of the following:

(a) The prediction and control of the spread of infectious diseases **2**

....

....

(b) The treatment of a non-infectious disease **2**

....

....

(c) The prevention of non-infectious disease **2**

....

....

(d) Assisting with the effects of a disorder **2**

....

....

(Sample question)

Question 32 (4 marks)

(a) PCR is a form of artificial DNA replication. What do the letters 'PCR' stand for? **1**

....

(b) Provide one example of how PCR is used and the benefits of its use. **3**

....

....

....

(Sample question)

Question 33 (4 marks) **Marks**

Explain how genes and the environment affect genotypes. **4**

(Sample question)

Question 34 (5 marks)

In your study you investigated the structure and function of proteins in living things.

(a) How is the shape or structure of the protein determined? **1**

(b) Name two different examples of proteins in living things and for each example, describe how the structure of the protein is related to its function. **4**

(Sample question)

Question 35 (5 marks) **Marks**

Outline the uses and advantages of TWO current genetic technologies. Construct a table for your answer. 5

(Sample question)

Question 36 (7 marks) **Marks**

Analyse the social implications and ethical uses of biotechnology. 7

(Sample question)

Sample HSC Examination Paper with 2017 HSC Questions

Sample Answers

Section I

1 C Homeostasis is the process of detecting and responding to changes in the internal environment of organisms.

2 D The nervous system uses sensory nerves to detect changes and responds by activating organs and muscles to respond to environmental changes.

3 A Robert Koch designed four rules of procedure for showing that a particular microorganism is the cause of a particular disease. The rules are called Koch's postulates.

4 A Lymphocytes include T and B cells and together contribute towards the immune response to fight infection.

5 C The grey line on the graph represents the hormone progesterone.

6 C Some white blood cells are called phagocytes because they engulf microorganisms.

7 A Renal dialysis replaces the role of the kidneys to remove toxic urea wastes from the blood.

8 C A polypeptide is a long chain of amino acids, which make up proteins. Peptide bonds join amino acids into the chain.

9 A The only difference between experiment Y and control W is the presence of the enzyme amylase, ensuring experimental validity.

10 B Insulin lowers blood glucose levels by facilitating glucose transport from the blood into cells. Glucagon raises blood glucose levels by metabolising glycogen into glucose.

11 C The surface of foreign cells contain antigens to which both T and B cell receptors respond.

12 B The heterozygous pea will produce half its gametes with the T allele and the other half with t while the homozygous short will only produce t which will be dominated by a T allele.

13 D In DNA A matches T and C matches G as bases on complementary strands.

14 C Cells of fungi have cell walls and protozoans are animal-like unicellular organisms without a cell wall.

15 D A colour different from either parent can result if both alleles are fully expressed.

16 D Germ-line mutations can be passed onto offspring and therefore alter the allele frequency in a gene pool. Somatic mutations can result in diseases but are not passed onto offspring. Mutagens are the causes of mutations.

17 A Mortality refers to the number of deaths in a population. Morbidity refers to the rate of disease in a population. Co-morbidity refers to the occurrence of two or more diseases together. Incidence refers to the number of new cases of a disease.

18 C Cell replication cannot occur without DNA replication and the unzipping of the two DNA strands.

19 A The gametes would carry the haploid number of chromosomes that would then combine together in the zygote of the zebronkey.

20 A Developing cultures of human stem cells is the only example provided of the future use and application of biotechnology. All other options are examples of past uses.

Section II

Question 21 (Total 4 marks)

(a) Two benefits are: the identification of behavioural or lifestyle risk factors and therefore better knowledge for treatment, management and prevention strategies; and breakthroughs in understanding genetic factors involved in the predisposition to and development of some cancers. *(2 marks)*

(b) An example of an epidemiological study is the link between Vitamin D deficiency during pregnancy and the associated adverse health outcomes in mothers, the developing baby, and children later in life. The benefits of this research are that women at high risk of having Vitamin D deficiency can be targeted, screened and treated with Vitamin D supplements. *(2 marks)*

Question 22 (Total 4 marks)

Type of disease	Name of disease	Cause of disease
Infectious disease	Measles	Airborne virus
Non-infectious disease	Atherosclerosis	Narrowing and hardening of the artery walls caused by smoking, high-fat diets, lack of exercise and stress

(4 marks)

Question 23 (Total 5 marks)

(a) Inflammatory response or inflammation response *(1 mark)*

(b) The inflammatory response isolates and destroys the pathogen and prepares the tissue for healing. It also provides a physical barrier to prevent further spread of the infection and make the host aware of the infection. *(2 marks)*

(c) One physical change: an increase in the diameter of blood vessels and a reduction in the velocity of blood flow, which allows blood flow to the area of infection to increase.

One chemical change: some chemicals activate endothelial cells to secrete proteins that bring about blood clotting; blood clots cut off the blood flow and help prevent the pathogen from entering the bloodstream and spreading to other parts of the body. *(2 marks)*

Question 24 (Total 7 marks)

(a) (i) Genotype: Aa, Bb, Gg (*1 mark*)

(ii)

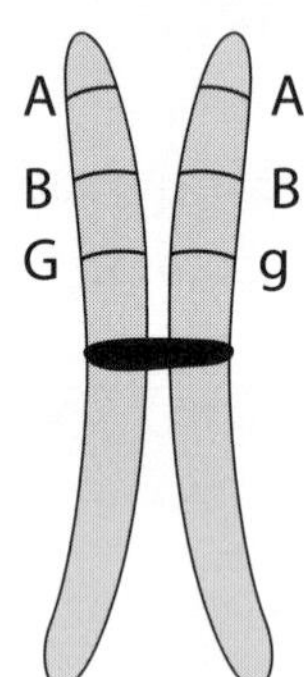

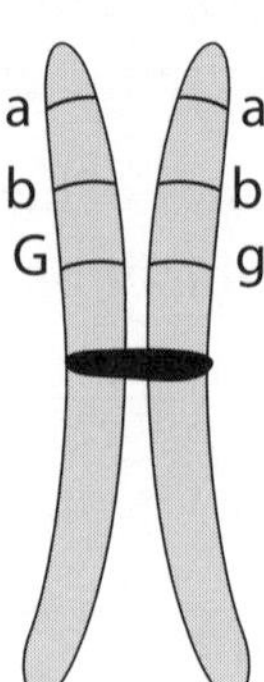

The middle two sequences are the recombined alleles resulting from crossing over. (*1 mark*)

(b) An independent assortment of chromosomes means that homologous pairs of chromosomes randomly segregate to opposite poles of the spindle during meiosis. The result is a large number of different chromosome combinations possible in each gamete. Because independent assortment happens in meiosis in both parents the genotype of the offspring could be highly varied. *(2 marks)*

(c) Crossing over produces chromosomes with new combinations of alleles. Gametes produced after crossing over has occurred will contain new combinations of maternally and paternally derived alleles. *(3 marks)*

Question 25 (Total 4 marks)

(a) Traditional knowledge is often exploited and Indigenous cultural and intellectual property rights are often not recognised. For example, Aboriginal people are concerned that their cultural knowledge of plants, animals and the environment is being used by scientists, medical researchers, nutritionists and pharmaceutical companies for commercial gain, often without Aboriginal peoples' informed consent and without any benefits flowing back to Indigenous communities. *(2 marks)*

(b) One process is the application of Aboriginal protocols in the development of some medicines and biological materials. Protocols serve as a guide for ethical and equitable benefit-sharing and management of research information and intellectual property. *(2 marks)*

Question 26 (Total 4 marks)

(a) The independent variable is substrate concentration. *(1 mark)*

(b) The trend in the graph is that initially an increase in substrate concentration increases the rate of reaction but after a certain concentration the rate of reaction remains unchanged. Enzymes catalyse biochemical reactions by lowering the activation energy because their active site combines with the substrate into an enzyme-substrate complex. The enzyme itself is used repeatedly with additional substrate molecules but each time it must bind with the substrate. Up to a certain increase in concentration of substrate molecules, more and more enzymes can bind and promote the reaction. Beyond that level other limiting factors come into play. It is possible at this point that the enzyme molecules are all combined with substrate molecules so no more are available to increase the rate of reaction. There may be the increase in concentration of products which might inhibit enzyme activity. *(3 marks)*

Question 27 (Total 5 marks)

(a) Prevention of infectious diseases refers to the range of measures that aim to reduce the risks of pathogens spreading or to stop infections in the first place. *(1 mark)*

(b) Effective hygiene practices prevent and control the spread of disease by reducing the chance of food becoming contaminated and infectious diseases being contracted. Practices include personal hygiene, such as washing hands, which helps to prevent the spread of such diseases as influenza. Public health campaigns are focused on prevention and control through strategies such as: health promotion information, focusing on behavioural risk factors relating to sexually transmitted diseases such as HIV/AIDS, intravenous drug use, and measures to reduce the risk of infection. *(4 marks)*

Question 28 (Total 3 marks)

This pattern of inheritance could be explained as sex-linked, i.e. carried on the X chromosome. For the female grandmother (1 in I) to be affected, she must be homozygous for the condition, i.e. have an allele for the condition on each of her X chromosomes. Supporting evidence is that both (100%) of her sons (3 and 5 in II) are also affected and one daughter (2 in II) is a carrier that is not affected but has one son (1 in III) that is affected. *(3 marks)*

Question 29 (Total 5 marks)

The diagnosis and treatment and management of diseases can extend and improve the lives of individuals who have established disease, but successful prevention of the disease itself and associated risk factors are required to reduce the incidence of non-infectious disease. Many non-infectious diseases such as cardiovascular diseases, cancer, diabetes and chronic respiratory diseases are the leading global cause of death (*Australia's Health Report*, 2016). Yet these diseases share key modifiable behavioural risk factors, such as tobacco use and an unhealthy diet. A fundamental aim of any healthcare system is to prevent disease so that people remain as healthy as possible, for as long as possible. An example is an increase in the prevention of lung cancer. The approach to this disease has benefited from epidemiology that has established a strong link between smoking as a highly correlated risk factor for contracting lung cancer, as well as some other cancers, and increased risks of atherosclerosis and strokes. Changes have occurred in legislation to restrict places where people can smoke and protect workers from second-hand cigarette smoke. Cigarette advertising has been banned, and anti-smoking public awareness and education campaigns have been conducted. Labelling on cigarette packets has also changed.

Justification for the change in emphasis is that there are far better health outcomes from prevention than treatment, both for society and individuals, as prevention is relatively easy, being about behavioural change. The change of emphasis does not imply that patients are no longer treated for lung cancer, surgically or with radio- or chemotherapy, and research is still ongoing to better target chemotherapy drugs, in particular, to cancer cells. *(5 marks)*

Question 30 (Total 6 marks)

Acquired immunity to a disease is achieved through the adaptive immune system and antibodies, B cells and T cells. It can confer long-term immunity. The adaptive immune system can remember the antigen because it produces memory cells. This is also the reason why there are some illnesses a person can acquire only once in their life, because afterwards their body becomes immune. After the first (primary) contact with the pathogen (antigen), it takes several days for the adaptive immune system to respond. After a second infection with the same pathogen, the adaptive immune system will recognise it and immediately release the antibodies needed to fight it. *(6 marks)*

Question 31 (Total 8 marks)

(a) Computer modelling that projects and simulates how a disease will progress and spread and inform public health interventions and treatments *(2 marks)*

(b) Proton therapy that delivers specific and focused radiation into a tumour damaging the DNA of the tumour cells *(2 marks)*

(c) Genetically modifying organisms such as bacteria and viruses to produce substances such as insulin that can help prevent disease *(2 marks)*

(d) Laser surgery that involves reshaping the cornea of the eye to correct refractive errors such as myopia *(2 marks)*

Question 32 (Total 4 marks)

(a) PCR stands for polymerase chain reaction. *(1 mark)*

(b) PCR is used in biotechnology to produce larger amounts of DNA from very small amounts of DNA or from impure samples. This is beneficial when larger amounts of DNA are needed; for example, in forensic analyses of samples of tissue from a crime scene. *(3 marks)*

Question 33 (Total 4 marks)

The total base composition and arrangement of the DNA in an organism is called its genome. The genetic composition is called the genotype. As diploid organisms have two copies of each gene, the genotype for a particular gene is the two forms of the gene or alleles that are present. The traits of an organism are qualities such as height and eye colour. The specific observable or measurable characteristics of an organism, such as blue eye colour or tall stature, is called its phenotype. The phenotype is the result of the interaction of the genotype and environment. An organism may have a genotype that should produce tall stature but malnutrition may mean the phenotype is only medium stature. *(4 marks)*

Question 34 (Total 5 marks)

(a) The shape or structure is determined by the sequence of amino acids on the polypeptide chain. *(1 mark)*

(b) Two examples are enzymes that promote and control chemical reactions and antibodies that detect harmful substances and organisms (antigens) such as bacteria.

Enzymes have specific structures at their active site that form the exact shape for the substrate molecule to bind to.

Antibodies are Y shaped. Antibodies recognise the specific molecule or antigen because of the structures on the tip of the Y. *(4 marks)*

Question 35 (Total 5 marks)

Technology	*Uses*	*Advantages*
Genetic engineering involves inserting genes from one organism into another (i.e. producing GMOs or recombinant DNA technology) Current technology includes gene editing tools such as the use of CRISPR-Cas9 and gene drives	Improves the sustainability and productivity of agriculture and protects plants, animals and humans from disease Produces medicines and pharmaceuticals, for example, hepatitis B vaccine Gene drives are currently being used in research to incorporate genes in order to prevent the ability of mosquitoes to carry the malaria parasite	New combinations of genes can dramatically improve the potential of agriculture, medicine, waste treatment and the pharmaceutical industry CRISPR-Cas9 has a target efficiency; that is, it does its job in the correct place on the DNA over 70% of the time
Gene silencing	RNAi technology can identify which genes are responsible for particular traits; plant breeders can then produce more productive but non-genetically modified plants	A therapeutic agent used to control disease and prevent infection in plant and animal cells. The process is fast, reliable, flexible, stable and controllable The ability to produce plants with desirable properties without using GM so RNAi is more acceptable to many people

(5 marks)

Question 36 (Total 7 marks)

Biotechnology is the use of living systems and organisms to make products. Common uses of biotechnology occur in agriculture (such as selective breeding and the development of crops with enhanced nutritional benefits), industry (such as fermentation process and the development of biofuels), environmental management (such as the development of biodegradable plastics and the development of products that can degrade toxic or harmful chemicals and wastes by the use of bacteria or other microorganisms) and medical uses and applications (such as the development of vaccines and antibiotics and the CRISPR-Cas9 gene editing tool). With the developments in biotechnology have come both social benefits and concerns including ethical implications. Social benefits include less starvation and under-nutrition, less environmental damage because of more efficient food production, less impact on communities because of drought resistant crops, increased life expectancy and quality of life, personalised medicine and better management of health risks, the elimination of some pathogens or vectors and more environmentally friendly and reliable energy supplies.

The social concerns including ethical implications include: concerns about the costs of aged care due to rapid improvements in life expectancy, particularly in developed countries; an increase in hormones and pesticides in food; biotechnologies inequitably distributed and often not available to many poorer people, with some arguing that the technologies widen the gap between wealthy, resource-rich communities and those who are less well-off; an increase in the influence on societies of companies involved in modern biotechnologies, the exclusive patents and control and high costs of these technologies; the possibility of irreversibly changing species and ecosystems through the use, for example, of GM crops and animals; the potential for bioterrorism or the use of biological weapons in warfare and concerns about future directions in biotechnology, such as the potential to create life or create designer babies.

(7 marks)

CHAPTER 8

Sample HSC Examination Paper with 2018 HSC Questions

Biology

General Instructions

- Reading time — 5 minutes
- Working time — 3 hours
- Write using black pen
- Draw diagrams using pencil
- NESA approved calculators may be used
- For questions in Section II, show all relevant working in questions involving calculations

Total marks: 100

Section I – 20 marks
- Attempt Questions 1–20
- Allow about 35 minutes for this section

Section II – 80 marks
- Attempt Questions 21–34
- Allow about 2 hours and 25 minutes for this section

Section I
20 marks
Attempt Questions 1–20
Allow about 35 minutes for this section

---

1 Which of the following is a known cause of non-infectious diseases in humans?

A. Diseases due to morbidity
B. Diseases caused by fungal pathogens
C. Diseases caused by environmental exposure
D. Diseases caused by viral pathogens

(Sample question)

2 In your study you investigated technologies that are used to assist with the effects of disorders in humans. Which technology in the table is correctly matched to the disorder it is designed to assist with the effects of?

	Technology	*Disorder*
A.	Laser surgery	Hearing loss
B.	Cochlear implants	Visual disorder
C.	Haemodialysis	Loss of kidney function
D.	Haemodialysis	Nutritional disease

(Sample question)

3 Which defence adaptation in the table is correctly matched with one of its features?

	Defence adaptation	*Feature*
A.	Inflammation	Constriction of blood vessels
B.	Phagocytosis	Production of antibodies by white blood cells
C.	Lymph system	Transportation of blood to help fight pathogens
D.	Cell death	Formation of barrier around the pathogen

(2018 HSC)

4 Which of the following desscribes the daughter cells produced as a result of mitosis?

A. Two genetically different cells

B. Two genetically identical cells

C. Four genetically different cells

D. Four genetically identical cells

(2018 HSC)

5 Which ONE of the following factors is most likely involved in limiting the spread of an infectious disease at the local level?

A. Border surveillance and protection

B. Travel bans

C. Screening of travellers

D. Public health and community-level measures

(Sample question)

6 Why is mRNA important in polypeptide synthesis?

A. It reads the genetic code to insert amino acids into the polypeptide chain.

B. It transfers the genetic code from DNA in the nucleus to the ribosome in the cytoplasm.

C. It leaves the nucleus and enters the cytoplasm.

D. It leaves the cytoplasm and enters the nucleus.

(Sample question)

7 Students dissected a mammalian kidney and drew the following diagram.

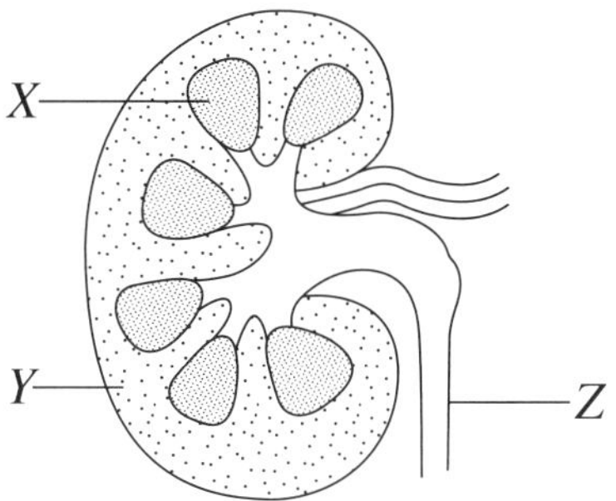

Which row in the table shows the correct labels for *X*, *Y* and *Z*?

	X	*Y*	*Z*
A.	Cortex	Medulla	Ureter
B.	Renal pyramid	Bowman's capsule	Renal artery
C.	Medulla	Cortex	Ureter
D.	Bowman's capsule	Cortex	Renal artery

(2018 HSC)

8 An organism suspected of causing a disease is described as being unicellular, having a cell wall and lacking a nucleus.

How is this organism classified?

A. A bacterium

B. A fungus

C. A protozoan

D. A virus

(2018 HSC)

9 Sunken stomata can be found in the leaves of some Australian plants. A section of such a leaf is shown.

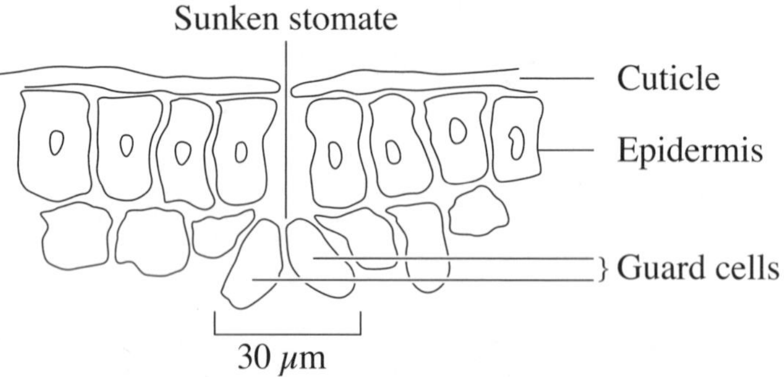

How do sunken stomata assist the plant to conserve water in a dry environment?

A. They trap moist air, reducing humidity.

B. They prevent entry of gases into the leaf.

C. They accumulate moist air, reducing transpiration.

D. They increase the surface area available for transpiration.

(2018 HSC)

10 Both artificial insemination and cloning are reproductive techniques that can decrease the genetic diversity of a population.

Which row of the table provides a correct reason for each technique's contribution to this decrease?

	Artificial insemination	*Cloning*
A.	Random fertilisation takes place	Large numbers of individuals are produced
B.	One male has many offspring	All gametes are genetically identical
C.	All male gametes are identical	All individuals have the same phenotype
D.	Fewer males are used to reproduce	All individuals have the same genotype

(2018 HSC)

11 The following plants were presented to a quarantine office in Australia as part of a shipment of plants entering Australia for the plant nursery trade.

Which of the following is a decision that the quarantine office is likely to make?

A. Plant *W* can enter Australia as it looks like it has 'black spot' which already occurs in Australia.

B. Plant *X* can enter Australia as it is unlikely the disease it has will transfer to Australian species.

C. Plant *Y* cannot enter Australia as it has a disease caused by shortage of soil magnesium.

D. Plant *Z* cannot enter Australia because its appearance suggests it may be carrying live insects.

(2018 HSC)

12 The graph shows four possible relationships between ambient temperature and body temperature.

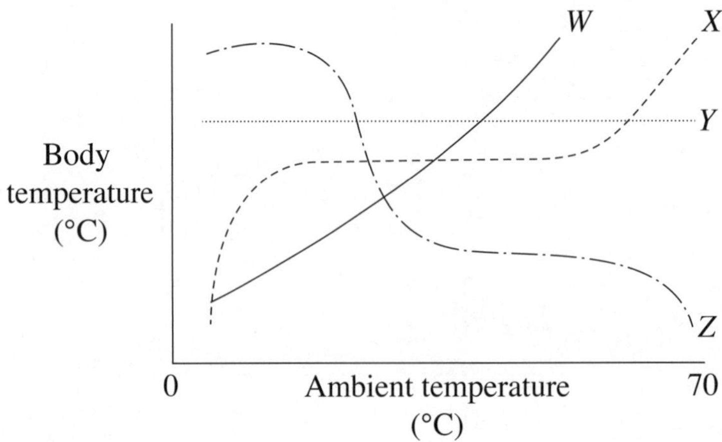

Which line on the graph represents the relationship between ambient temperature and body temperature for an endotherm in a terrestrial environment?

A. W

B. X

C. Y

D. Z

(2018 HSC)

13 Bt cotton is a type of cotton now commonly grown that contains a gene from the soil bacterium *Bacillus thuringiensis.* As in the bacterium, the gene in the Bt cotton produces a protein that is toxic to insect larvae. The benefit of this technology is the reduced need for pesticides.

This is an example of:

A. a genetically modified organism.

B. whole-organism cloning.

C. propagation.

D. artificial selection.

(Sample question)

14 The following pedigree shows the inheritance of a disorder.

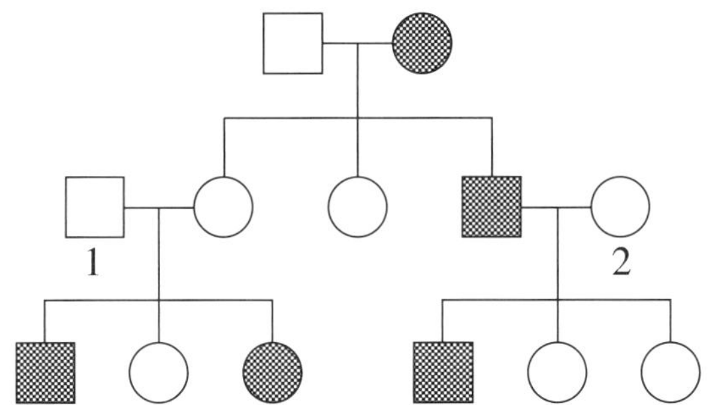

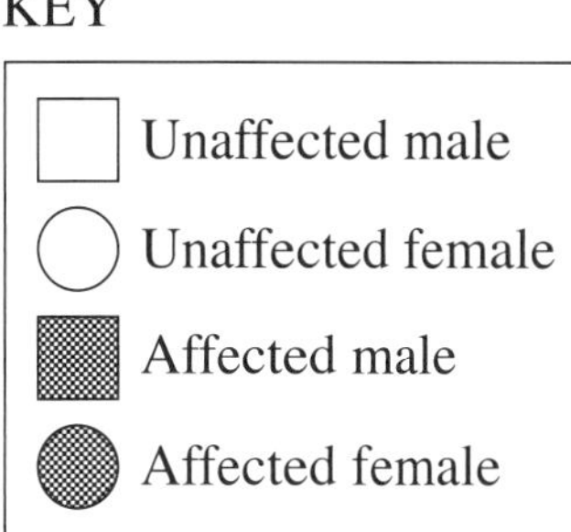

Which row of the table shows the genotypes of individuals 1 and 2?

	Individual 1	*Individual 2*
A.	Aa	Aa
B.	AA	Aa
C.	X^AY	X^AX^a
D.	X^aY	X^AX^a

(2018 HSC)

15 Which row of the table is correct with respect to the substances that pass through the nephron?

	Bowman's capsule	*Proximal convoluted tubule*	*Collecting duct*
A.	Chloride ions are at the same concentration as in plasma	All chloride ions are reabsorbed	Water is reabsorbed in the presence of aldosterone
B.	Plasma proteins are at the same concentration as in plasma	Water is reabsorbed by osmosis	Water is reabsorbed in the presence of ADH
C.	Glucose is at the same concentration as in plasma	All glucose is reabsorbed	Water is reabsorbed in the presence of aldosterone
D.	Sodium ions are at the same concentration as in plasma	Sodium ions are reabsorbed	Water is reabsorbed in the presence of ADH

(2018 HSC)

16 How do helper T cells assist in raising a specific immune response to a pathogen?

A. They mass produce specific antibodies.

B. They stimulate the cloning of specific T cells.

C. They are cloned and differentiate to become specific cytotoxic T cells.

D. They produce cytokines that stimulate the cloning of specific phagocytes.

(2018 HSC)

17 The diagram shows the life cycle of the malaria parasite, *Plasmodium sp*. Five stages in this life cycle are numbered on the diagram.

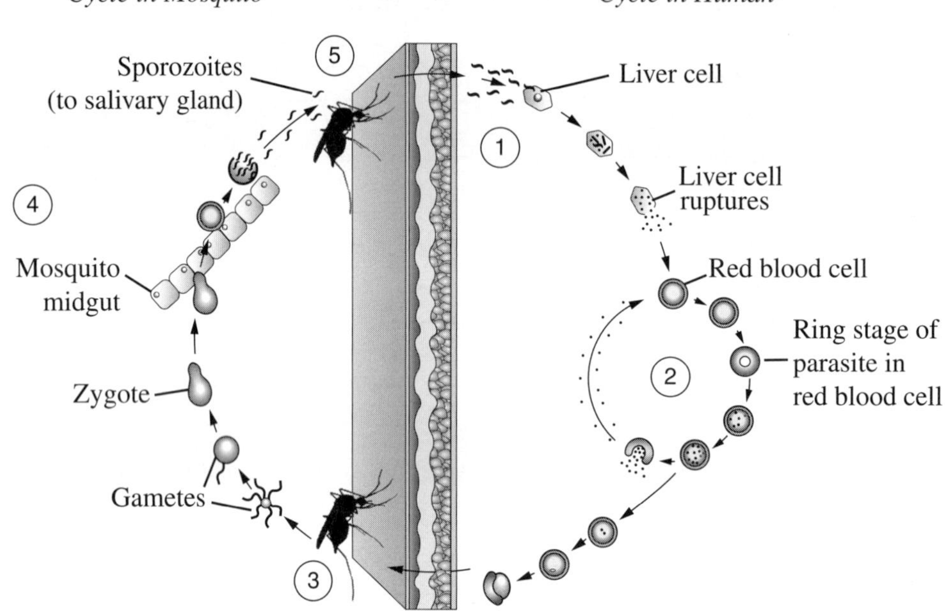

To prevent malaria, the following four strategies have been used:

- taking anti-malarial drugs
- spraying swamps with insecticides
- using mosquito nets over beds
- administering a malaria vaccine.

Which row in the table shows the stage in the life cycle in which each of these strategies would be most effective?

	Taking anti-malarial drugs	*Spraying swamps with insecticides*	*Using mosquito nets over beds*	*Administering a malaria vaccine*
A.	5	1	3	2
B.	2	5	1	3
C.	1	4	2	5
D.	2	3	5	1

(2018 HSC)

18 In your study you evaluated the historical, culturally diverse and current strategies used to predict the spread of infectious disease. Which of the following is an example of a current strategy?

A. Genetic screening

B. Integrated global data and surveillance

C. Pharmaceuticals

D. Data from healthcare centres

(Sample question)

19 Colour blindness in humans is determined by a sex-linked gene. Two family trees are shown.

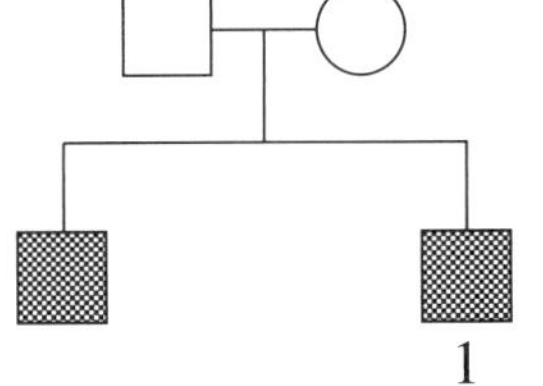

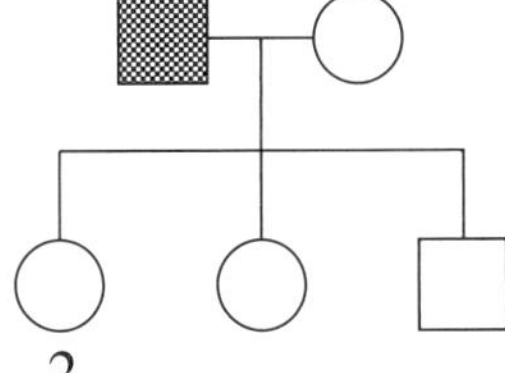

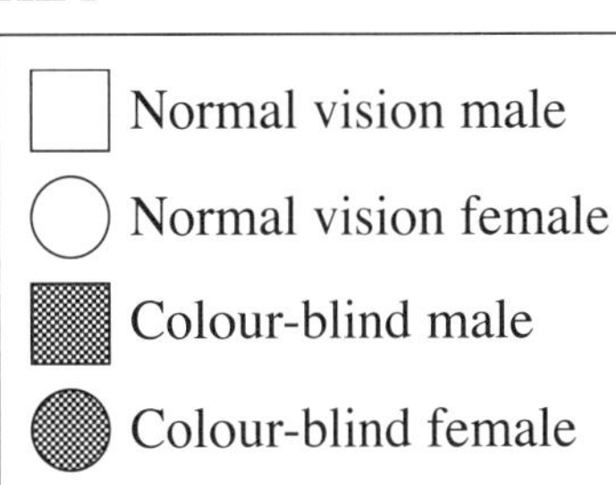

Which row of the table shows the probability of colour-blind offspring of each sex if individuals 1 and 2 were to have children together?

	Male offspring	*Female offspring*
A.	0%	0%
B.	50%	0%
C.	50%	50%
D.	0%	0%

(2018 HSC)

20 The table below provides data on the number of new cases of three cancers diagnosed in a population in a year. What does this data represent?

Type of cancer	*Number of new cases diagnosed in a year*
Breast cancer	17 586
Melanoma of the skin	13 941
Lung cancer	11 556

A. Survival rates

B. Mortality rates

C. Prevalence rates

D. Incidence rates

(Sample question)

Biology

Section II
Answer Booklet

80 marks
Attempt Questions 21–34
Allow about 2 hours and 25 minutes for this section

Instructions

- Answer the questions in the spaces provided. These spaces provide guidance for the expected length of response.
- Show all relevant working in questions involving calculations.

Please turn over

Question 21 (4 marks) **Marks**

(a) Identify TWO responses of a named endotherm to a decrease in body temperature. **2**

....

....

....

(b) Outline the role of the nervous system of an endotherm in maintaining homeostasis when its body temperature changes. **2**

....

....

....

....

....

Question 22 (5 marks)

(a) Pasteur performed an experiment to identify the role of microbes in decay. **2**

Justify a conclusion that can be drawn from his results.

....

....

....

....

(b) Describe the contribution of Robert Koch to our understanding of disease. **3**

....

....

....

....

....

....

(2018 HSC)

Question 23 (5 marks) **Marks**

(a) A student plans to conduct a practical investigation relating to microbial testing of water or food samples. **3**

Complete the table, identifying features that this student should include to ensure valid experimental design.

Dependent variable	•
Control	•
Variables to be kept constant	• •

(b) Describe TWO safety precautions that the student could undertake to minimise risks associated with this practical investigation. **2**

....

....

(Sample question)

Question 24 (6 marks) **Marks**

A new flu vaccine is prepared each year to protect the population against the current strains of influenza virus. The effectiveness of flu vaccines varies from year to year and can be measured using the overall vaccination effectiveness (VE) index. A VE of 60% means that a vaccinated individual's chance of getting the flu is reduced by 60%.

The following data show the VE over a 10-year period.

Influenza season	*VE (%)*
2006–2007	52
2008–2009	41
2010–2011	60
2012–2013	49
2014–2015	19

(a) Draw an appropriate graph to represent the data on the following grid. **3**

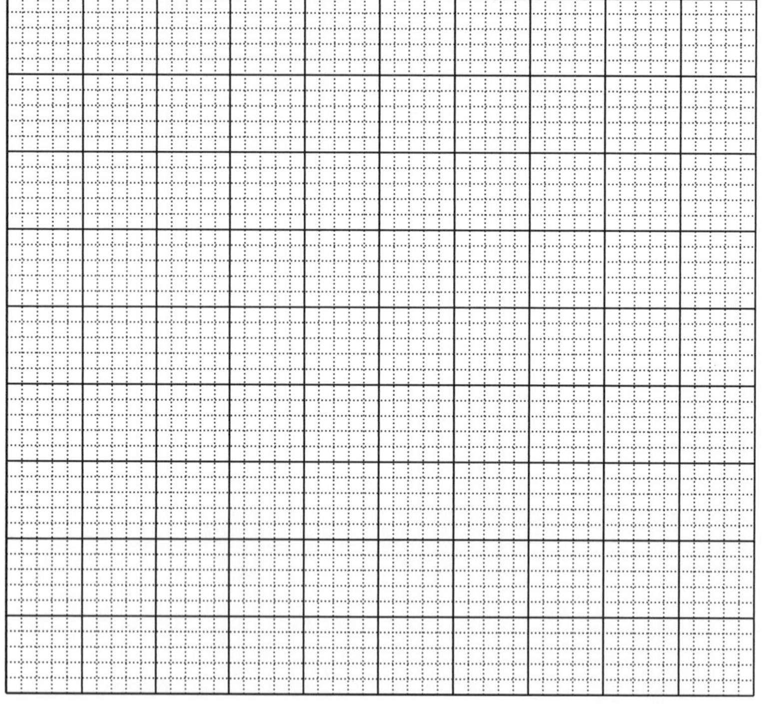

Question 24 continues

Question 24 (continued) **Marks**

(b) Provide a possible explanation for the vaccination effectiveness (VE) index in 2014–2015. **3**

(2018 HSC)

End of Question 24

Question 25 (4 marks)

Construct a table to compare the processes and outcomes of two reproductive technologies: artificial insemination and artificial pollination. **4**

(Sample question)

Question 26 (6 marks) **Marks**

(a) Compare proteins and polypeptides. **2**

....

....

....

....

(b) Antidiuretic hormone (ADH) is a protein produced by cells in the hypothalamus. The AVP gene codes for the production of ADH.

(i) Outline the steps to show how a mutation in the AVP gene could result in changes in the ADH protein. **3**

....

....

....

....

....

....

....

....

(ii) Identify ONE possible effect of the AVP mutation on kidney function. **1**

....

....

(2018 HSC)

Question 27 (6 marks) **Marks**

(a) Define the terms *genotype* and *phenotype*. **2**

....

....

....

(b) A population of mammals living in a very cold climate have long, thick fur which assists their survival. At times, a widespread infestation of mites causes their fur to become thin, and bald patches appear in their coats. In this population some individuals are resistant to the mites. This is due to the presence of a co-dominant allele (M^R). Individuals that are homozygous for this allele (M^RM^R) are infertile. **4**

Explain how both genotype and phenotype influence the inheritance of genes and natural selection in this population.

....

....

....

....

....

....

....

....

....

....

....

....

....

....

....

....

....

(2018 HSC)

Question 28 (7 marks) **Marks**

The diagram models the process of meiosis.

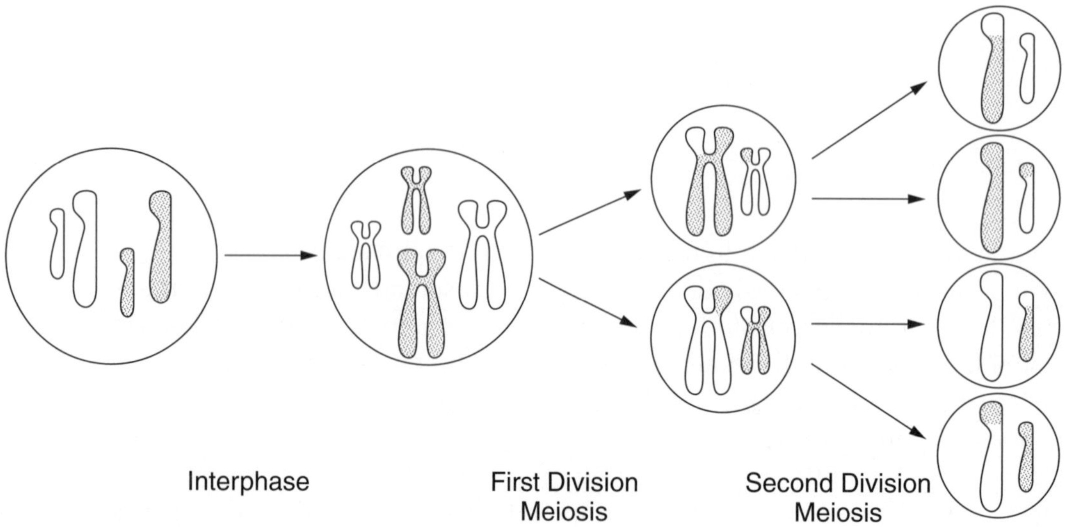

Interphase

First Division Meiosis

Second Division Meiosis

(a) Describe the process that accounts for the changes shown in the model during interphase. **2**

....

....

....

....

Question 28 continues

Question 28 (continued) **Marks**

(b) Explain the structure and behaviour of chromosomes in the first division of meiosis. Include detailed reference to the model. **5**

....

....

....

....

....

....

....

....

....

....

....

....

....

....

....

....

....

(2018 HSC)

End of Question 28

Question 29 (8 marks) **Marks**

The graph shows the expected life span (the age to which people are expected to live in years) for people of different ages during the 20th century in one country. **8**

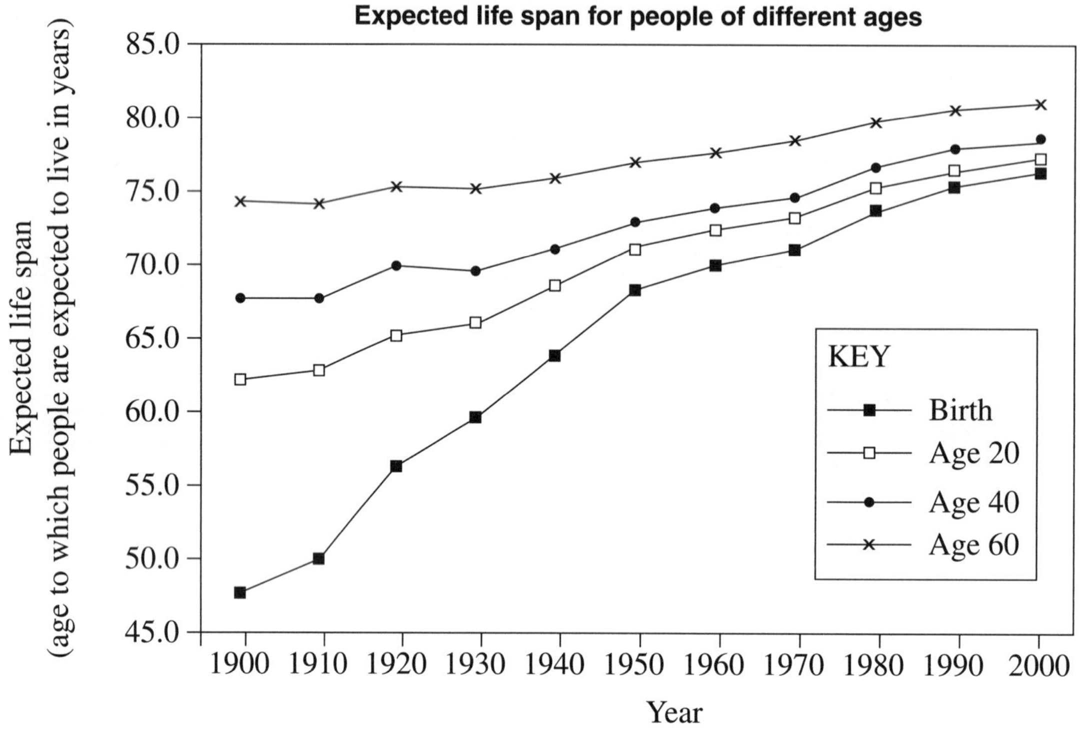

Expected life span for people of different ages

Expected life span (age to which people are expected to live in years)

85.0
80.0
75.0
70.0
65.0
60.0
55.0
50.0
45.0

1900 1910 1920 1930 1940 1950 1960 1970 1980 1990 2000

Year

KEY
Birth
Age 20
Age 40
Age 60

There have been many biological developments that have contributed to our understanding of the identification, treatment and prevention of disease.

Evaluate the impact of these developments on the expected life span. In your answer, include reference to trends in the data provided.

....

....

....

....

....

....

....

....

....

Question 29 continues

Question 29 (continued)

(2018 HSC)

End of Question 29

Question 30 (9 marks) **Marks**

(a) (i) Identify TWO structures that refract light as it passes through the eye. **2**

(ii) The diagram shows the human eye. The two lines labelled *A* and *B* repesent two beams of light passing through the eye and stimulating different areas of the retina. **2**

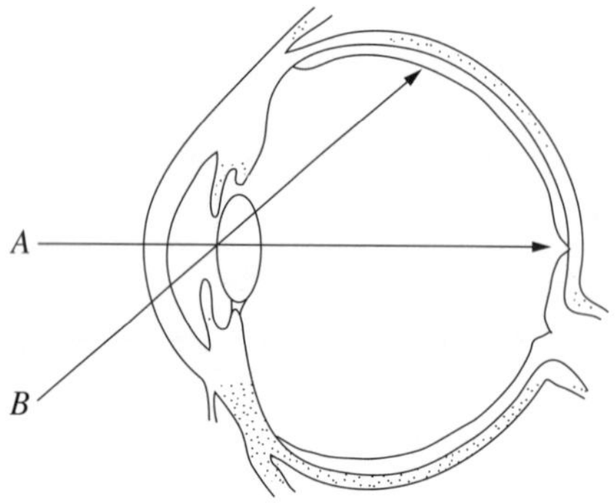

A

B

The visual perception will be different at the two stimulated areas.

Identify TWO of these differences.

(2018 HSC)

(b) A person with normal vision and a person with myopia are both looking at an object in the distance. **5**

Construct THREE labelled diagrams of an eye to show the light path through the eye of the person with:

- normal vision
- myopia
- myopia corrected with a suitable lens.

(2018 HSC)

Question 31 (5 marks) **Marks**

Data from Australia's Health Tracker 2017 provides information on factors that increase the risk of some non-infectious diseases. For one non-infectious disease prevalent in Australia:

(a) name the disease and list TWO risk factors that can cause it. **2**

....

....

(b) describe the treatment and/or management of the disease. **1**

....

....

(c) explain how epidemiological studies can be of benefit in the treatment and management of the disease. **2**

....

....

....

(Sample question)

Question 32 (4 marks)

Diversity in a gene pool may result from factors such as population size and diverse environments. List and describe TWO other factors that can affect the gene pool of populations, especially small populations. **4**

....

....

....

....

(Sample question)

Question 33 (5 marks) **Marks**

'Immunisation is an important public health intervention to stop the spread of diseases that can cause serious illness and death. Worldwide immunisation programs prevent an estimated 3 milllion deaths every year.' (Australian Institute of Health and Welfare, 2016)

(a) What is immunisation? **1**

....

....

(b) Provide TWO examples of how immunisation has prevented disease world-wide and in Australia. **2**

....

....

(c) Describe TWO other public-health interventions that help stop the spread of disease. **2**

....

....

(Sample question)

Question 34 (6 marks) **Marks**

'Globalisation appears to be causing profound and sometimes unpredictable changes in conditions that shape the burden of diseases in some populations. Evidence suggests that changes in these conditions have led to changes in prevalence, spread and geographical range and control of many infectious diseases, particularly those transmitted by vectors. **6**

Discuss this statement.

....

....

....

....

....

....

....

....

....

....

....

(Sample question)

Sample HSC Examination Paper with 2018 HSC Questions

Sample Answers

Section I

1 C Some non-infectious diseases, such as lung and skin cancer, are known to be caused by environmental exposure. Morbidity refers to the rate of disease in a population. Fungal and viral pathogens cause infectious diseases.

2 C Haemodialysis is an artificial way of removing waste substances from the blood, which is a function usually performed by the kidneys.

3 D Cell death is a defence adaptation that can seal off an infected area.

4 B Mitosis is cell replication resulting in two genetically identical cells.

5 D Public health and community-level measures are implemented locally to limit the spread of infectious diseases. Other factors such as border protection are interrelated.

6 B mRNA transfers the genetic code from DNA in the nucleus to the ribosome in the cytoplasm. It does leave the nucleus and enter the cytoplasm but this option doesn't explain its importance. tRNA reads the genetic code to insert amino acids into the polypeptide chain.

7 C The outer layer of the kidney is the cortex that contains the glomeruli.

8 A Prokaryotic bacteria lack nuclei but can have cell walls and be unicellular.

9 C Sunken stomata reduce air flow around the point of transpiration.

10 D Artificial insemination increases the reproductive potential of a small number of males and cloning results in offspring genetically identical to the nuclear donor parent and each other.

11 D Quarantine attempts to prevent the entry of pathogens.

12 B Endotherms maintain a relatively constant body temperature over a narrow range of temperature change.

13 A Bt cotton is an example of gene technology or genetic engineering that results in genetically modified organisms (GMO).

14 A The condition is autosomal recessive. Both individuals are unaffected but carry one allele for the condition.

15 D Filtration in Bowman's capsule results in only larger molecules remaining in the blood and then active reabsorption of sodium ions occurs in the proximal distal tubule. ADH is a hormone that increases the permeability of the collecting ducts so that water can be reabsorbed.

16 D Helper T cells release cytokines and help activate macrophages.

17 D Prevention of malaria by spraying swamps stops mosquito vectors from breeding and the nets prevent transfer of sporozoites from the mosquito saliva into humans.

18 B The question asks for current strategies used to **predict** the spread of infectious disease. Current strategies use computer technologies for fast processing of data gathered globally. Data from health-care facilities is still used but is more traditional. Genetic screening is a prevention strategy. Pharmaceuticals are often treatment strategies.

19 C A colour-blind male and a female carrier of the condition have a 50% chance that either male or female offspring will have the condition.

20 D Incidence is the number of new cases diagnosed in a given period of time.

Section II

Question 21 (Total 4 marks)

(a) Humans are endotherms that in response to a decrease in body temperature may experience vasoconstriction in the region of the skin and increase metabolic heat production by shivering. *(2 marks)*

(b) The nervous system is responsible for thermoreceptors detecting the change in body temperature and then through the central nervous system activating effector organs such as the muscles that commence shivering or control the constriction or dilation of blood vessels. *(2 marks)*

Question 22 (Total 5 marks)

(a) A conclusion that can be drawn from the results of Pasteur's swan-neck flask experiment is that microbes do not spontaneously generate but are a result of direct contact with the air. The conclusion is justified as the flask that was exposed to air became contaminated whereas the one that was sealed from the air by the convoluted shape of the neck of the flask remained uncontaminated. *(2 marks)*

(b) Koch developed rules of procedure (Koch's postulates) that if followed would demonstrate a direct causal link between the presence of a pathogen and the disease it causes. The steps involved developing pure cultures from infected organisms and infecting heathy organisms with the pure culture and comparing the cultures once the exposed organisms became infected. He identified the bacterial causes of diseases such as anthrax and tuberculosis. *(3 marks)*

Question 23 (Total 5 marks)

(a) Dependent variables are the variables that change in response to the independent variables.

Dependent variable	• The types of colonies that grew from the water or food sample
Control	• An agar petri dish on which no sample was grown but which was treated in exactly the same way otherwise; that is, not inoculated
Variables to be kept constant	• The same amount of agar medium • The same temperature for each petri dish

(3 marks)

(b) Autoclaving of agar plates following incubation and the use of protective clothing, gloves and eyewear *(2 marks)*

Question 24 (Total 6 marks)

(a)

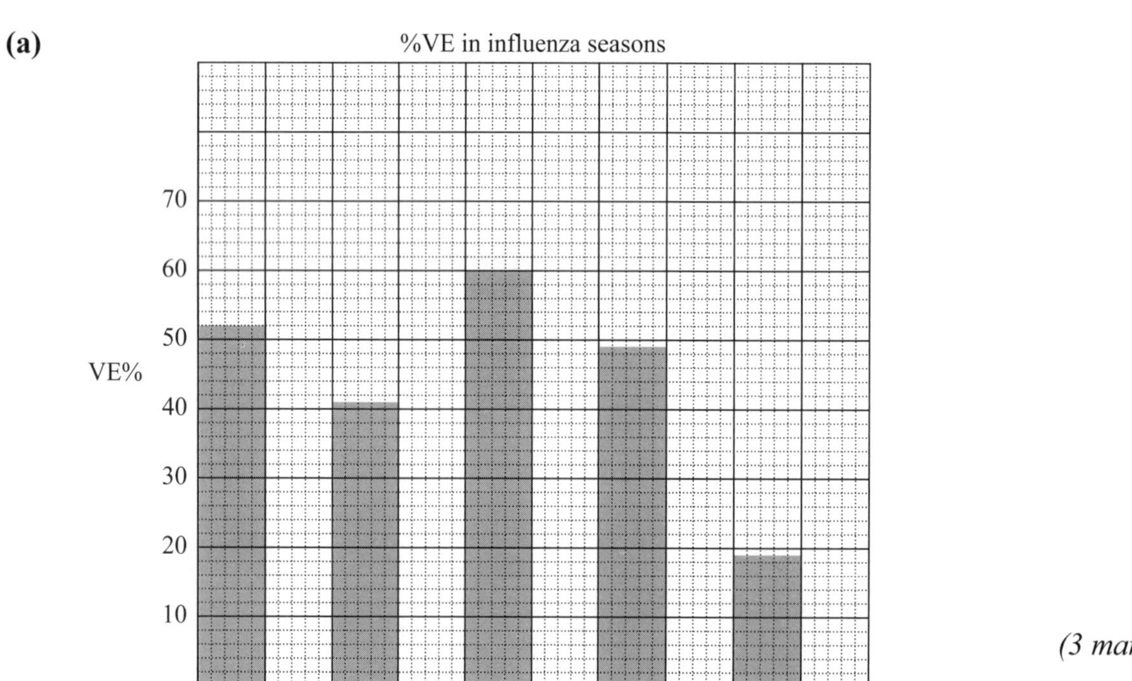

(3 marks)

(b) In 2014–15 an individual's chance of contracting the influenza as a result of having the flu vaccine was only reduced by 19%. This was possibly the result of the large number of mutations that occur in viruses such as the flu virus so that many strains survive simultaneously. The vaccines are developed using a specific strain and may be ineffective against other strains. Because of the highly contagious nature of flu viruses, the strains that remain virulent spread widely or quickly. The effectiveness of a vaccine is dependent on the vaccine promoting the development of artificially acquired immunity. *(3 marks)*

Question 25 (Total 4 marks)

Process and outcomes of two reproductive technologies

Reproductive technology	*Processes*	*Outcomes*
Artificial insemination	Collection, storage and transport of semen from male animals with desirable qualities to fertilise a large number of females	Larger numbers of more desirable animals Global benefits from storage and transport of semen
Artificial pollination	Involves the transfer of pollen to stigmas to enable pollination and fertilisation	Improved plant cultivars such as plant hybrids with more vigour and therefore increased productivity beyond that of either parent

(4 marks)

Question 26 (Total 6 marks)

(a) A polypeptide is a chain of specific amino acids joined together by peptide bonds. Secondary and tertiary polypeptide structures result from folding and twisting of the amino-acid chain. A protein consists of at least one polypeptide chain that has a specific conformation. The shape of the combination of polypeptide chains (called quaternary structure) is the result of the interacting forces between the chains. The resultant protein structure is significant for its function. *(2 marks)*

(b) (i) A mutation in the AVP gene could result in:

- a substitution of a nitrogenous base in the DNA that could result in a different amino acid in the polypeptide chain that is part of the ADH protein. This may affect the structure and function of the protein or may prematurely terminate the process of polypeptide synthesis
- a deletion or insertion of a DNA base that results in a frameshift mutation that greatly impacts on AVP formation and its ability to function as a hormone.

(3 marks)

(ii) An AVP mutation could affect the ability of the kidney to reabsorb water and hence the water balance in the blood. *(1 mark)*

Question 27 (Total 6 marks)

(a) The term *genotype* means the genetic make-up. It may refer to the make-up of a pair of specific alleles for a specific characteristic. The term *phenotype* refers to the manifestation of the genotype as a physical or physiological trait, such as eye colour in humans or blood type. The phenotype results from the expression of the genotype and the interaction with the environment of the organism. *(2 marks)*

(b) Codominance results from the interaction of two different alleles that are both expressed in the heterozygous form. In the case of resistance to the mite the heterozygous mammals will have the advantage of not developing bald patches and having a phenotype resulting in them being better able to resist the cold climate. Heterozygous individuals will pass on to half their offspring the M^R allele, thus continuing the allele frequency in the population. Homozygous individuals for the mite-resistant gene are not able to pass on the allele as they have the phenotype that results in them being infertile, tending to reduce the frequency of the allele that conveys resistance to mites. Natural selection will favour the heterozygous phenotype and genotype and maintain a certain level of allele frequency in the population. *(4 marks)*

Question 28 (Total 7 marks)

(a) The model shows the chromosomes replicating during interphase. Each chromosome as a result consists of two identical chromatids joined by a centromere. Before this occurs the DNA that is an important component of chromosomes must replicate. This means the double helix unwinds, the strands separate and new strands build using the original strand as a template. The replicated chromosomes contain replicated DNA and hence can carry the genetic material to the next generation. *(2 marks)*

(b) Key events in the first division of meiosis following the formation of identical chromatids include the crossing over of adjacent chromatids from homologous chromosomes. As a result of crossing over, new combinations of alleles are formed on some of the chromatids that eventually separate to become chromosomes. Another process in the first meiosis division is the alignment of the homologous pairs in a random assortment. When the spindles contract, drawing the homologous pairs of chromosomes to opposite poles (and hence different cells), new combinations of chromosomes result. The other key event in the first division is the reduction of the chromosome number from the diploid to the haploid number. This allows gametes to form with the haploid chromosome number, and when they return to the diploid number the potential for genetic variation increases even further. *(5 marks)*

Question 29 (Total 8 marks)

The graph indicates that in the country shown life expectancy for people born in 1900 was relatively low (about 48 years), but in that year 20-year-olds could expect to live until 65 years and 60-year-olds could expect to live a further 14–15 years, to about 75. This suggests a high infant mortality rate. The data suggests that a high incidence of infectious diseases reduced life expectancy, but once exposure to pathogens in people with healthy immune systems had resulted in naturally acquired immunity, people were able to live relatively long lives. A relatively active lifestyle, little global travel, and nutrition that was not based on 'fast foods' and sugary drinks might also have contributed to a reasonably high life expectancy for those surviving childhood and adolescence.

In about 1950 in most countries improved hygiene, better food preservation and refrigeration, improved treatment in hospitals, the recent introduction of antibiotics and the expansion of vaccination programs resulted in dramatic improvements in the life expectancy of newborn babies. Some improvement in life expectancy for 20-, 40- and 60-year-olds had also occurred with an increasing understanding of viruses such as poliovirus, resulting in improvements in treating infectious diseases. Non-infectious diseases were to become of increasing concern. This time was also marked by the new awareness of the molecular basis of inheritance and so the beginning of the possibilities of understanding and treating inherited disorders.

Improvements in life expectancy since the 1950s has remained almost parallel for the four different age groups, with only a slightly better improvement for newborn babies. These improvements could be the result of recently improved understanding of inherited disorders that has led to earlier (even prenatal) diagnosis and screening. Widespread use of antibiotics has been impeded by the development of resistant strains of bacteria. New drugs for the treatment of viruses and better vaccines and vaccination programs have contributed to life expectancy improvements. Expansion of worldwide travel has resulted in some epidemics such as SARS becoming pandemics. This has been countered by attempts at better quarantine. Non-infectious diseases such as Type II diabetes, strokes and ischaemic heart disease have become the focus of public-health campaigns to reduce obesity and improve lifestyle. Environmental exposure to carcinogens has led to changes in regulations about smoking and drinking and 'slip, slop, slap' campaigns. Genetic engineering now has the potential to reduce the spread of vector-borne diseases such as malaria although up until 2000 the use of pesticides was an important prevention method. New technologies have improved the treatment of disorders such as kidney failure and the development of artificial body parts such as heart valves has improved quality and quantity of life. *(8 marks)*

Question 30 (Total 9 marks)

(a) (i) The cornea and the lens refract light as it passes through the eye. *(2 marks)*

(ii) The light passing horizontally through the eye (*A*) interacts at the fovea in the central macula of the retina. This area contains a high concentration of cones that are responsible for colour vision and visual acuity. Light ray *B* strikes the peripheral region of the retina that contains a high proportion of rods that detect shape and movement and discriminate between shades of light and dark in low-light intensities. *(2 marks)*

(b) Normal vision involves accommodation. The shape of the lens changes to result in incoming light focusing on the retina. The first diagram shows normal vision of a distant object.

The second diagram shows myopia (the eyeball is too long so the image is blurred) and the third shows correction using a concave lens that lengthens the distance at which an image is formed on the retina.

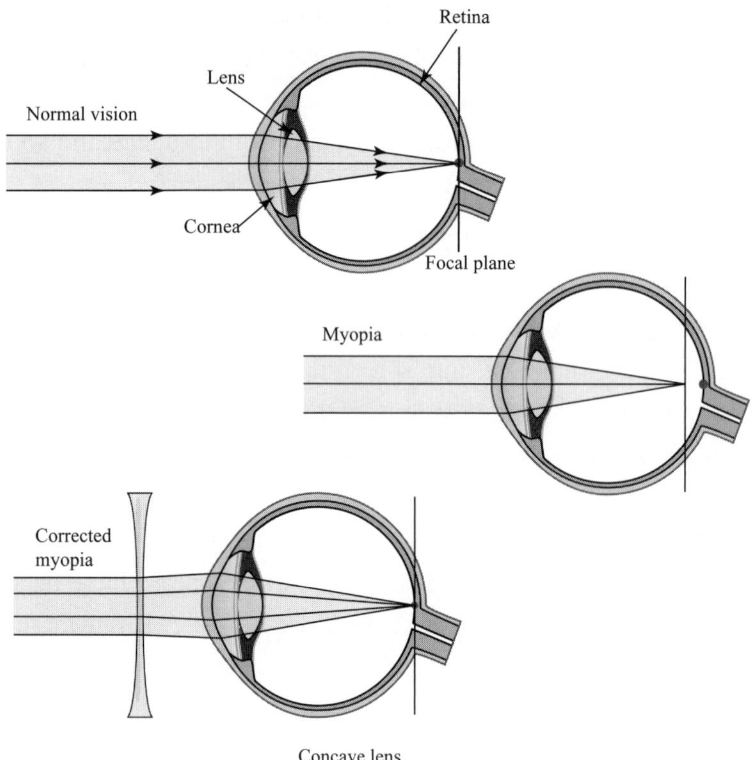

(5 marks)

Question 31 (Total 5 marks)

(a) Type 2 diabetes is a nutritional disease in which the body becomes resistant to the normal effects of insulin and/or gradually loses the ability to produce enough insulin in the pancreas. Two risk factors that can lead to Type 2 diabetes include insufficient physical activity and high saturated fat intake. *(2 marks)*

(b) These risk factors are modifiable; therefore management and prevention include increased levels of physical activity and changes to the diet. *(1 mark)*

(c) Epidemiology is the study of the patterns and causes of disorders and diseases in a defined population. The studies help identify trends in data such as changes over time. For example, studies that show a correlation between overweight and obesity and the increased risk of having Type 2 diabetes provide data to help prevent and manage the disease. Data from epidemiological studies has reduced the incidence of certain diseases such as lung cancer resulting from smoking. *(2 marks)*

Question 32 (Total 4 marks)

Mutations and gene flow affect the gene pool of populations. The effect is much greater in smaller populations. (You could have also given the answer as gene drift.)

Mutations are changes in DNA that increase the variety of alleles in the gene pool.

Gene flow is the movement of genes between populations due to migration. This results in genetic diversity. *(4 marks)*

Question 33 (Total 5 marks)

(a) Immunisation is the process of using vaccines to protect against illnesses caused by infection. Immunisation means both receiving a vaccine and becoming immune to a disease post-vaccination. *(1 mark)*

(b) The effectiveness of vaccines in preventing infectious diseases is shown by the disappearance, or near disappearance, in Australia of deaths from, for example, diphtheria and worldwide from polio. *(2 marks)*

(c) Two other public-health interventions are quarantine that prevents pathogens from coming into the state or country and public-health campaigns that focus on prevention and control through strategies such as health-promotion information and disease-prevention programs. *(2 marks)*

Question 34 (Total 6 marks)

The incidence (the number of new cases) and prevalence (the number or proportion of cases) of infectious diseases is influenced by a wide range of factors that includes the mobility of individuals and the proportion of the population that are immune or immunised. For most of human history regional and global populations have been isolated from one another. New infectious diseases could only spread as fast or as far as people could walk. Wars, exploration and trade have brought greater population movement and mobility of individuals and therefore the spread of diseases into populations that had evolved with relatively small gene pools along with no exposure to particular diseases and therefore no immunity. Now the speed and volume of mobility of individuals and human populations is unprecedented in human history. Vector-transmitted diseases, such as malaria, are now able to spread further and faster. Increased mobility through migration and tourism, for example, increases the contact between humans and vectors, bringing humans who are more susceptible or have no immunity to vector-borne diseases to areas where vectors thrive. New arrivals, refugees, displaced populations, and legal and illegal workers can bring disease vectors and drug-resistant strains, such as anti-malarial drug resistance, into areas where they were not present.

Changes in geographical distribution of vector-transmitted diseases such as malaria and dengue fever occurs as a result of population movement, population growth and rapid urbanisation. *(6 marks)*

CHAPTER 9

NSW Education Standards Authority

2019 HIGHER SCHOOL CERTIFICATE EXAMINATION

Biology

General Instructions

- Reading time – 5 minutes
- Working time – 3 hours
- Write using black pe
- Draw diagrams usin
- Calculators approved by NESA may be used

Total marks: 100

Section I – 20 marks (

- Attempt Questions 1–20
- Allow about 35 minutes for this section

Section II – 80 marks (pages 13–32)

- Attempt Questions 21–33
- Allow about 2 hours and 25 minutes for this section

Section I

20 marks
Attempt Questions 1–20
Allow about 35 minutes for this section

Use the multiple-choice answer sheet for Questions 1–20.

---

1 Which of the following is an example of a non-infectious disease?

A. Polio caused by a virus

B. Cholera caused by a bacterium

C. Wheat rust caused by a fungus

D. Haemophilia caused by a gene mutation

2 What does the body produce in response to a vaccine?

A. Antigens

B. Antibiotics

C. Antibodies

D. Activated toxins

3 The diagram shows the impact of birds feeding on a population of beetles over time.

Time

Which of the following accounts for the change in the beetle population?

A. Mutation

B. Gene flow

C. Genetic drift

D. Environmental pressure

4 The figure shows a part of a DNA molecule.

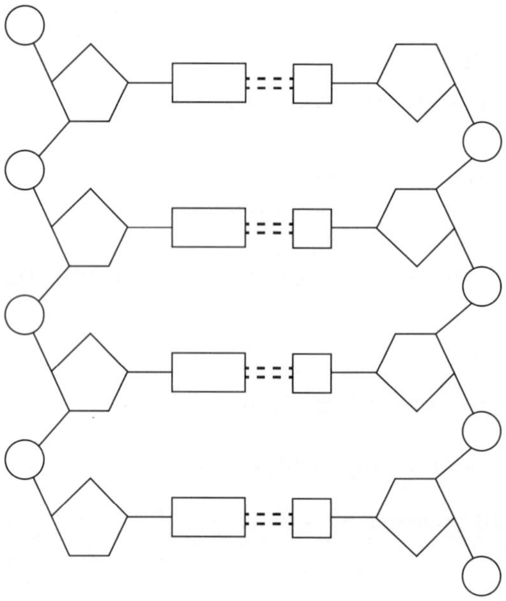

How many complete nucleotides are shown in the figure?

A. 1

B. 2

C. 4

D. 8

5 Which of the following is part of the innate immune response?

A. Antibodies

B. Phagocytes

C. Stomach acid

D. B lymphocytes

6 How does the cochlear implant assist people with severe hearing loss?

A. It amplifies sound.

B. It stimulates the ear drum.

C. It stimulates the auditory nerve.

D. It amplifies vibrations in the cochlea.

7 Two types of bacteria were isolated from a patient's throat swab and grown in pure culture on separate agar plates. On each plate there were FOUR different antibiotic discs, *W*, *X*, *Y* and *Z*.

The photograph shows the plates seven days later.

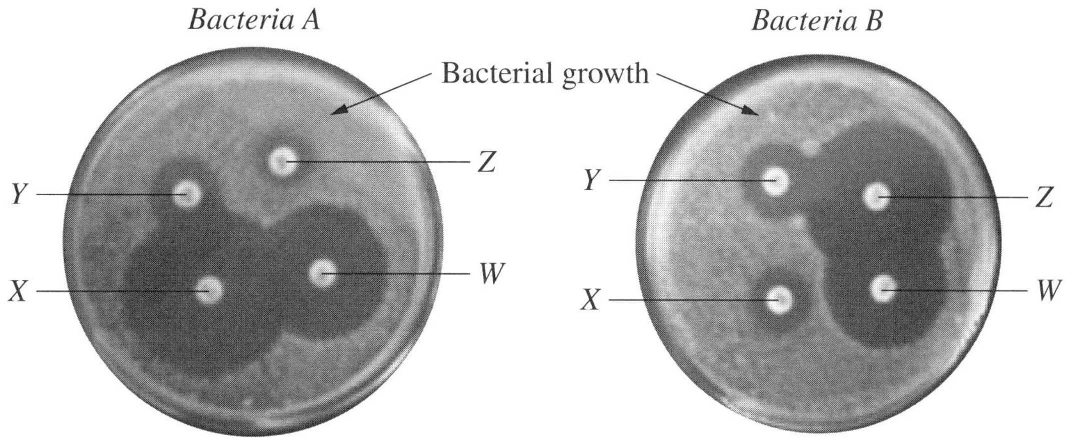

Which antibiotic should be used to treat the patient?

A. *W*

B. *X*

C. *Y*

D. *Z*

8 A group of four students set out to determine the animal species diversity over an area of one hectare in each of five different habitats. Each student graphed their data as shown.

Which of the graphs produced is the most suitable to represent animal species diversity in the different habitats?

A.

Number of animals

Rainforest
Wet tropical forest
Dry tropical forest
Salt marsh
Dry plains

B.

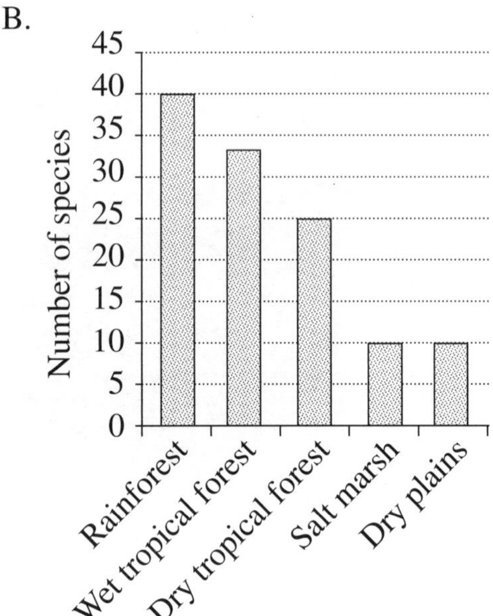

C.

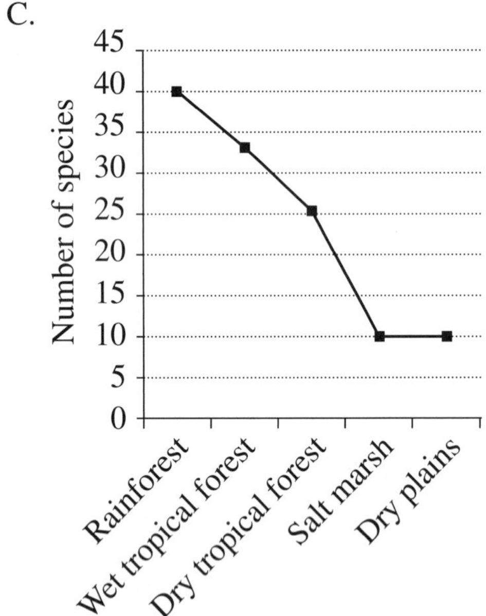

D.

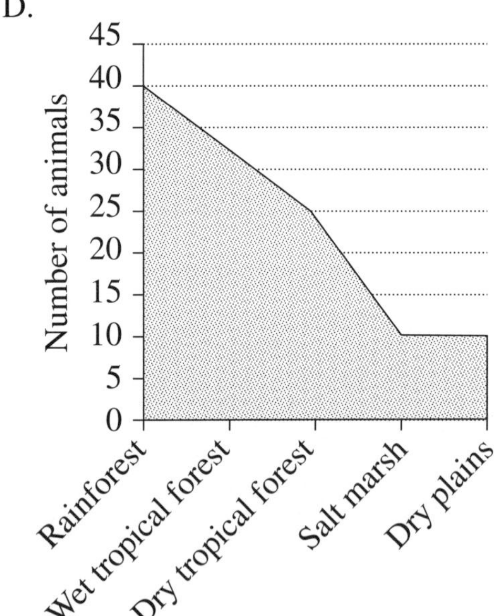

9 Which of the following is an advantage for animals using internal fertilisation rather than external fertilisation?

A. It prevents dehydration of gametes.

B. It involves large numbers of gametes.

C. It relies on adaptations such as mating rituals.

D. It allows gametes to combine to form unique offspring.

10 A group of islands are separated from each other by large stretches of water. Each island has its own policy on quarantine.

A nearby country is experiencing an outbreak of an infectious disease in its cattle.

An investigation is to be designed to find which of the quarantine policies operating on the islands is the most effective.

Which of the following would be a suitable design feature of the investigation?

A. The control is the smallest island.

B. The control is the number of infected cattle.

C. The independent variable is quarantine policy.

D. The independent variable is the number of infected cattle.

11 Which of the following is always true of a mutation that produces a dominant allele?

A. It will be lethal in a population.

B. It will be expressed in heterozygous individuals.

C. It will only be expressed in homozygous individuals.

D. It will spread more quickly through the population than a recessive allele.

12 The glucose tolerance test is used to investigate the control of glucose in the human body. Patients consume 75 g of glucose and their blood glucose is monitored.

Type 2 diabetes is a condition where the cells of the body do not respond adequately to insulin.

Which graph could represent the results of glucose tolerance tests in a non-diabetic person and a person with untreated Type 2 diabetes?

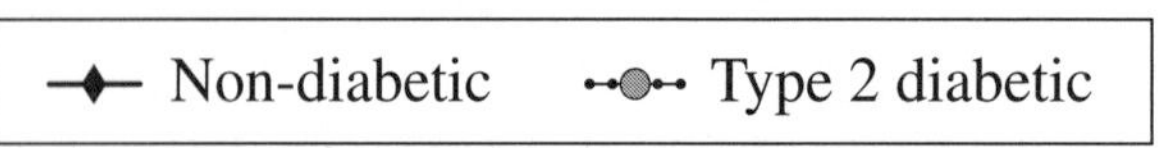

A.

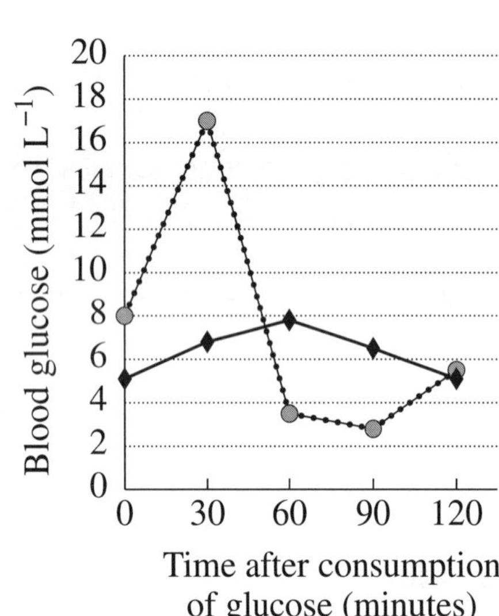

B.

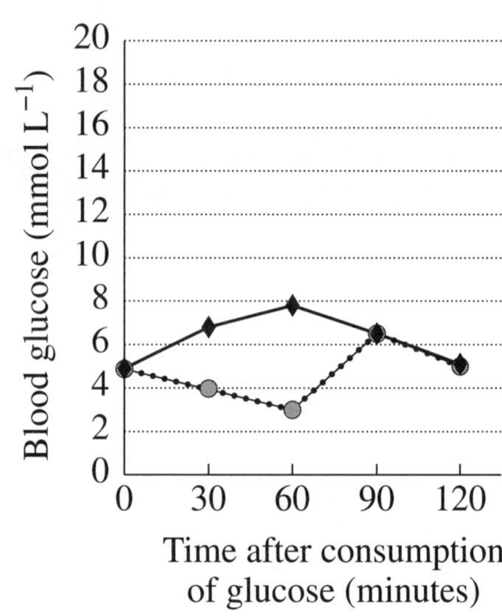

C.

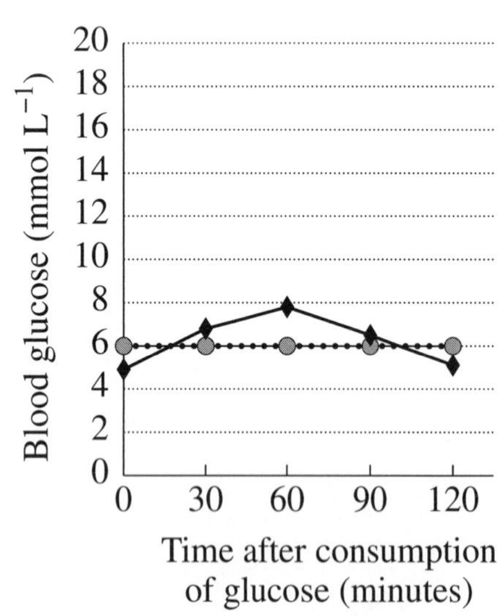

D.

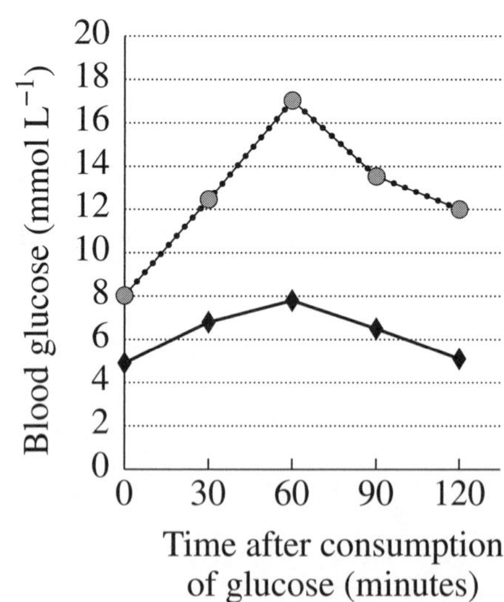

13 Genetic drift is a gradual change in

A. the alleles of an individual due to mutation.

B. allele frequency in a population due to chance.

C. the genes of a population due to natural selection.

D. gene frequency in a population due to natural selection.

14 The following DNA base sequence is used to code for a sequence of four amino acids.

CGC ATC ATG CTA

Which of the following correctly represents the anticodons on the transfer RNA during synthesis of this string of amino acids?

A. GCG UAG UAC GAU

B. CGC AUC AUG CUA

C. CGC ATC ATG CTA

D. GCG TAG TAC GAT

15 A germ-line mutation is known to have occurred.

How is it possible that there has been no noticeable change in the phenotype of the offspring?

A. The mutation occurred in a stretch of RNA.

B. The mutation occurred in a protein-coding region.

C. The mutation occurred in a stretch of non-coding DNA.

D. The mutation did not affect the DNA sequence of any gametes.

16 The diagram shows the concentration of an antibody to a particular pathogen.

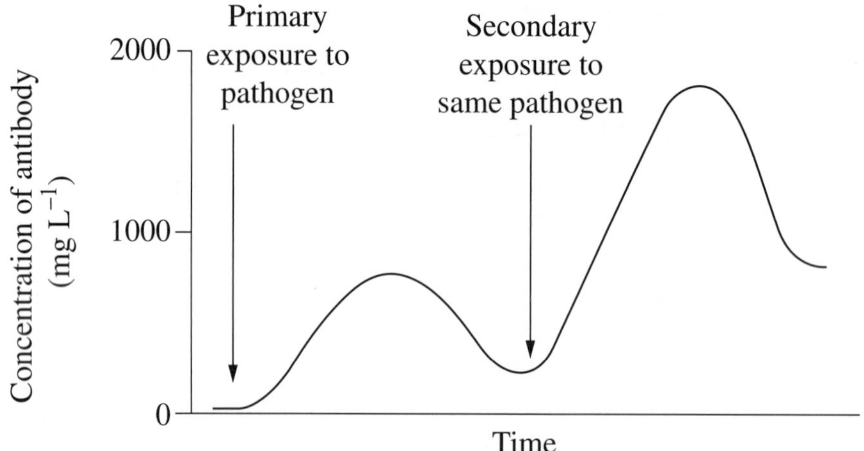

Two students are exposed to the pathogen. Student *X* had previously been vaccinated against this pathogen. Student *Y* had never been exposed to it.

Which row of the table shows the most likely levels of antibody in the blood of each student a week after exposure to the pathogen on this occasion?

Concentration of antibody in the blood ($mg\ L^{-1}$)

	Student X	*Student Y*
A.	250	1500
B.	1500	1000
C.	1000	1500
D.	1500	250

17 The pedigree shows the inheritance of a genetic disorder.

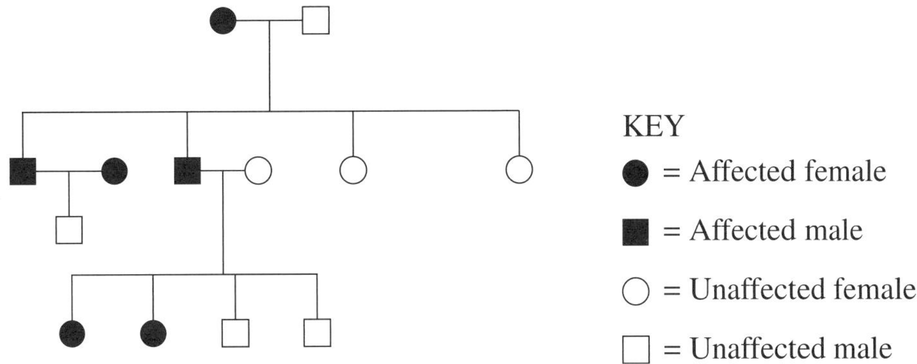

Which row of the table correctly identifies the two possible types of inheritance for this disorder?

	Autosomal dominant	*Autosomal recessive*	*Sex-linked dominant*	*Sex-linked recessive*
A.	✓		✓	
B.	✓			✓
C.		✓	✓	
D.		✓		✓

18 The diagram shows the effect of the hormone oxytocin on the uterus during the birth of a mammal.

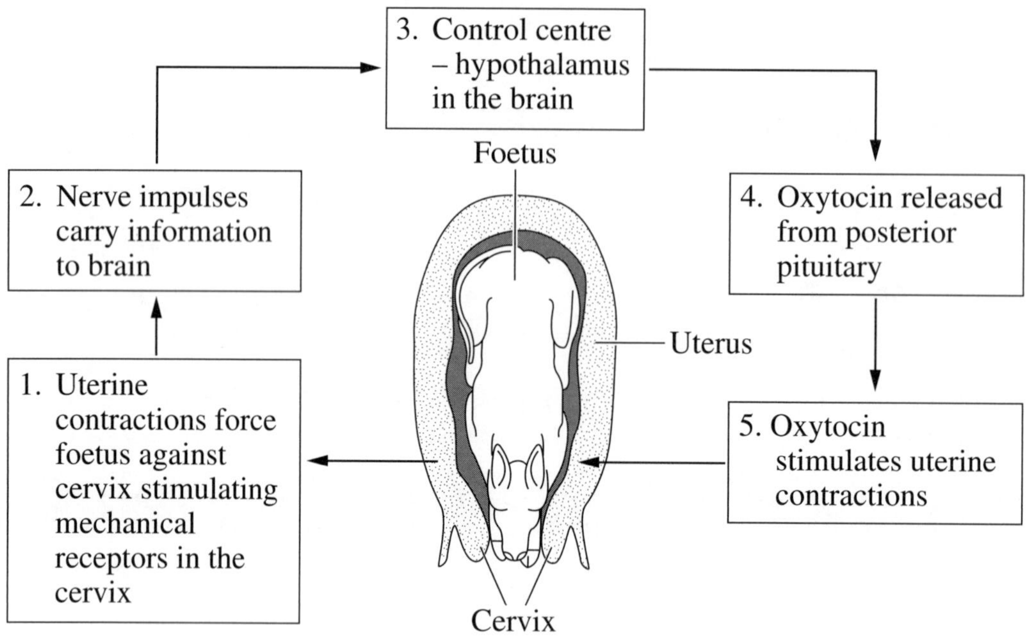

Which of the following best identifies and explains the feedback loop demonstrated in the diagram?

	Feedback loop	*Explanation*
A.	Negative	The production of oxytocin results in the production of more oxytocin.
B.	Positive	The production of oxytocin results in the production of more oxytocin.
C.	Negative	The production of oxytocin results in the detection of the contraction by receptors in the cervix.
D.	Positive	The production of oxytocin results in the detection of the contraction by receptors in the cervix.

Use the following diagram to answer Questions 19–20.

The diagram shows how CRISPR/Cas9 can be used as a new tool for genetic engineering. This technology has dramatically improved scientists' ability to successfully modify genomes.

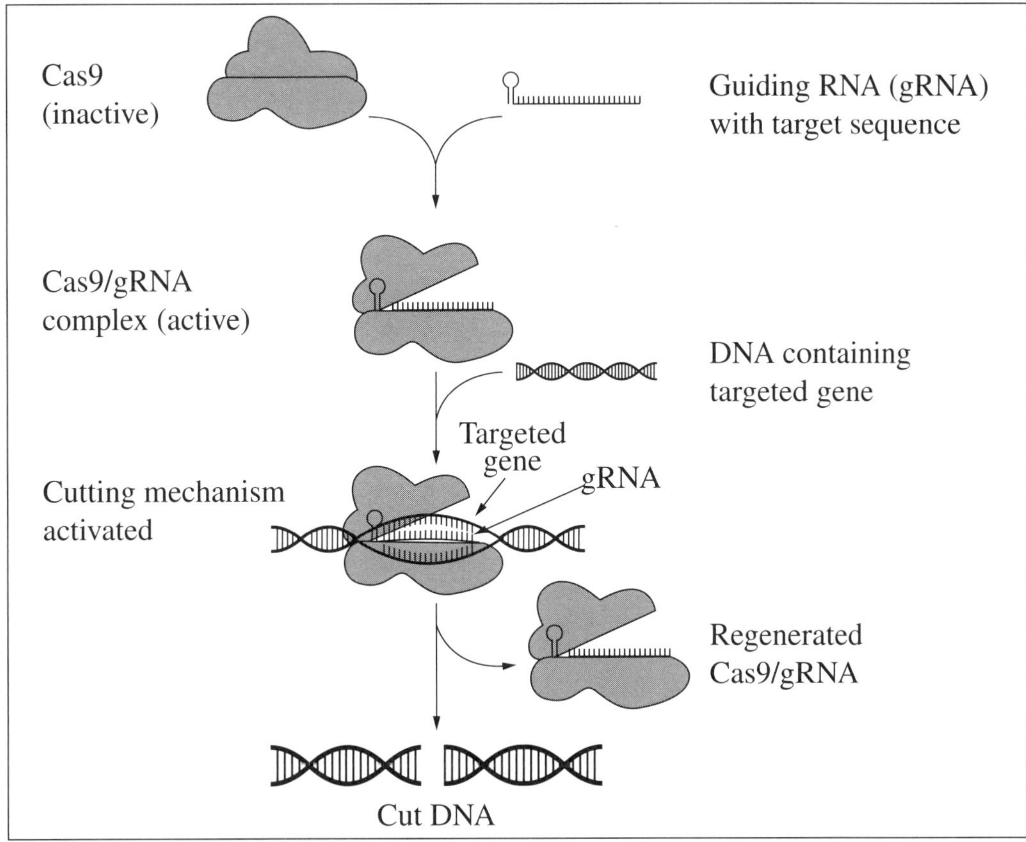

19 What type of structure must Cas9 be?

A. Enzyme

B. mRNA

C. Ribosome

D. tRNA

20 Scientists have been able to use biotechnology to 'cut and paste' DNA for decades.

Why would the new CRISPR/Cas9 technology have improved the scientists' success in cutting DNA of specific genes?

A. Cas9 is able to combine with specific DNA.

B. Cas9 has an active site that cuts target DNA.

C. gRNA has the same nucleotides as the target DNA.

D. gRNA has nucleotides complementary to the target DNA.

2019 HIGHER SCHOOL CERTIFICATE EXAMINATION

Centre Number

Student Number

Biology

Section II Answer Booklet

80 marks
Attempt Questions 21–33
Allow about 2 hours and 25 minutes for this section

Instructions

- Write your Centre Number and Student Number at the top of this page.
- Answer the questions in the spaces provided. These spaces provide guidance for the expected length of response.
- Show all relevant working in questions involving calculations.

Please turn over

Question 21 (3 marks)

The diagram shows a flow chart of the reaction of a human body to an increase in temperature. **3**

Fill in the three blank steps on the flow chart.

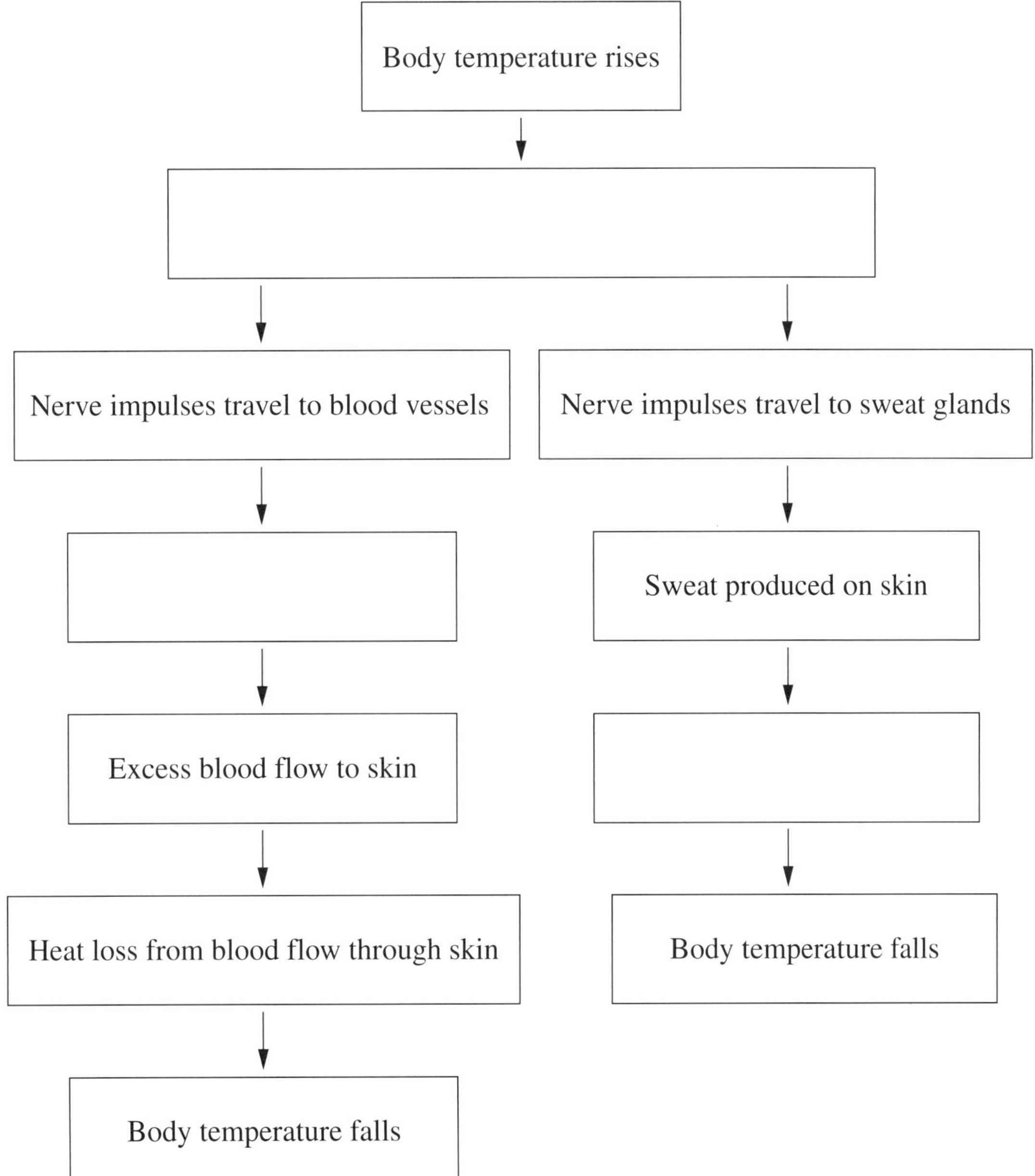

Question 22 (3 marks)

Complete the table to show the differences between *somatic* and *germ-line mutations*. 3

	Somatic mutation	*Germ-line mutation*
Location		
Effect on offspring		
Example		

Question 23 (5 marks)

Explain how educational programs can be effective in reducing the incidence of non-infectious diseases. Support your answer with examples. 5

....

....

....

....

....

....

....

....

....

....

....

....

....

....

....

Question 24 (5 marks)

Explain the loss of biodiversity that may result from TWO biotechnologies used in agriculture. **5**

Question 25 (5 marks)

A human karyotype that shows evidence of chromosomal mutation is shown.

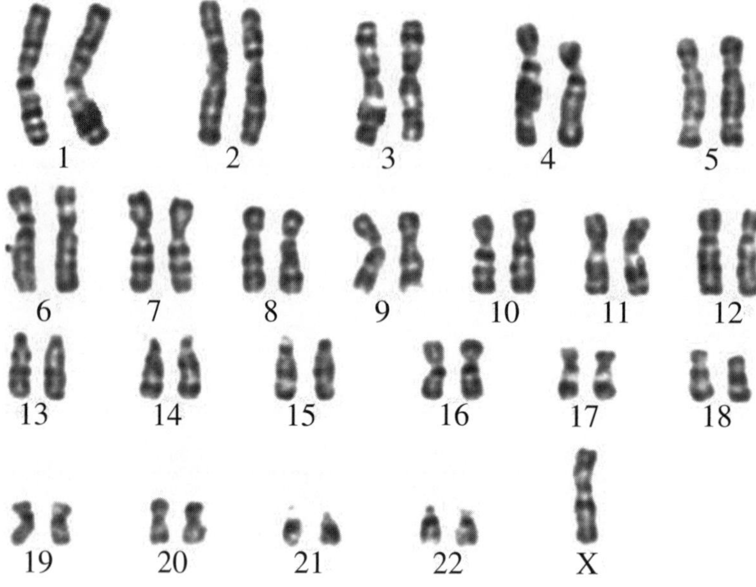

(a) Identify the evidence of chromosomal mutation in the karyotype. **1**

....

....

(b) Explain how cell division and fertilisation could lead to the production of this karyotype. **4**

....

....

....

....

....

....

....

....

....

....

Question 26 (5 marks)

The map shows the percentage of adult indigenous populations able to digest lactose. **5**

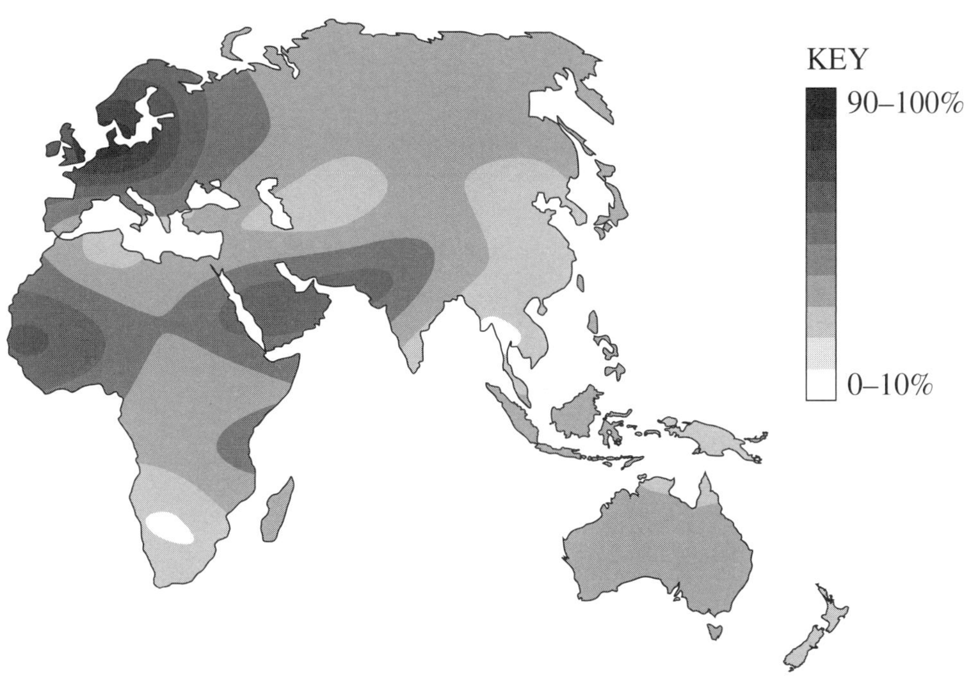

The ability to digest lactose is due to the presence of an enzyme (lactase) which can metabolise the sugar (lactose) present in milk. The gene responsible for producing lactase is usually permanently switched off at some time between the ages of 2 and 5 years. However, some people remain able to digest lactose throughout their lives.

With reference to evolution and DNA, provide possible reasons for the distribution shown in the map.

....

....

....

....

....

....

....

....

....

....

....

....

Question 27 (5 marks)

Yeast is a single-celled fungus that can reproduce by *budding*.

(a) What type of reproduction is *budding*? **1**

....

(b) Outline a procedure that could be used to test the effect of temperature on reproduction in yeast. **4**

....

....

....

....

....

....

....

....

....

....

....

....

Please turn over

Question 28 (6 marks)

Huntington's disease is an autosomal dominant condition caused by a mutation of a gene on chromosome 4. It causes nerve cells to break down.

Stargardt disease is an autosomal recessive condition caused by a mutation of a different gene on chromosome 4. It causes damage to the retina.

A patient is heterozygous for both Huntington's (Hh) and Stargardt disease (Rr). His father's extended family has numerous cases of both of these diseases. His mother does not have either disease and is homozygous for both genes.

(a) Complete the tables, showing the TWO alleles the patient inherited from each parent. **2**

Alleles from father

Alleles from mother

(b) The diagram shows the patient's homologous pair of chromosome 4 at various stages of meiosis. **4**

Add the relevant alleles to the diagram to model the production of possible gamete combinations. Include a key and an example of crossing over.

Homologous pair of chromosome 4 before crossing over	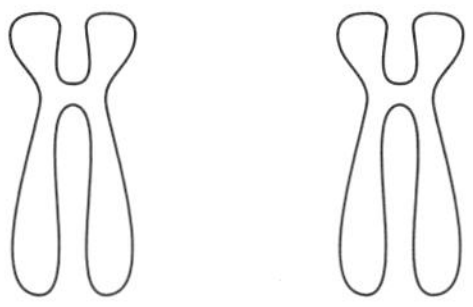
Homologous chromosomes after crossing over and separation	
Gametes	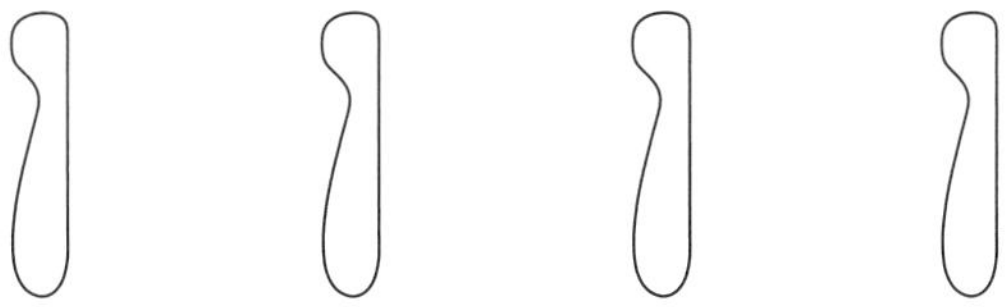

KEY

Question 29 (3 marks)

Describe ONE mechanism by which plants maintain internal water homeostasis. **3**

....

....

....

....

....

....

....

....

....

....

Please turn over

Question 30 (5 marks)

Experiments were conducted to obtain data on the traits 'seed shape' in plants and 'feather colour' in chickens. In each case, the original parents were pure breeding and produced the first generation (F1). The frequency data diagrams below relate to the second generation offspring (F2), produced when the F1 generations were bred together. **5**

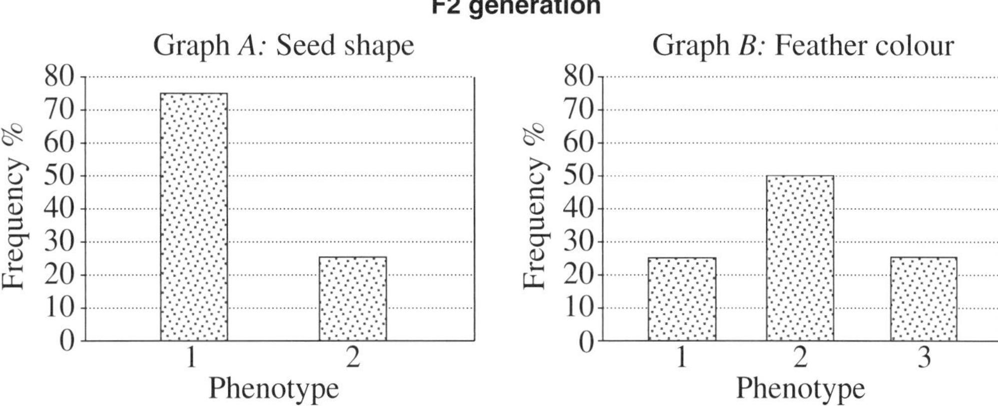

Explain the phenotypic ratios of the F2 generation in both the plant and chicken breeding experiments. Include Punnett squares and a key to support your answer.

Question 31 (5 marks)

(a) Outline ONE adaptation of a specific pathogen that facilitates its entry into a host. **2**

....

....

....

(b) Explain how the mode of transmission of pathogens influences the spread of diseases. **3**

....

....

....

....

....

....

....

....

....

Please turn over

Question 32 (10 marks)

Use the following data to answer parts (a) and (b).

Dengue fever and malaria are examples of infectious diseases transmitted between humans by mosquitoes. Dengue fever is caused by a virus transmitted by mosquitoes of the genus *Aedes.* Malaria is caused by a single-celled organism transmitted by mosquitoes of the genus *Anopheles*.

The following data provide information about the global incidence of these two diseases over time.

Global malaria data for selected years from 1900 to 2010

Year	*Global population* ($\times 10^9$)	*Number of countries with reported cases*	*Population at risk*	
			($\times 10^9$)	(%)
1900	1.2	140	0.9	75
1946	2.4	130	1.6	67
1965	3.4	103	1.9	65
1975	4.1	91	2.1	51
1992	5.4	88	2.6	48
1994	5.6	87	2.6	46
2002	6.2	88	3.0	48
2010	6.8	88	3.4	50

Question 32 continues

Question 32 (continued)

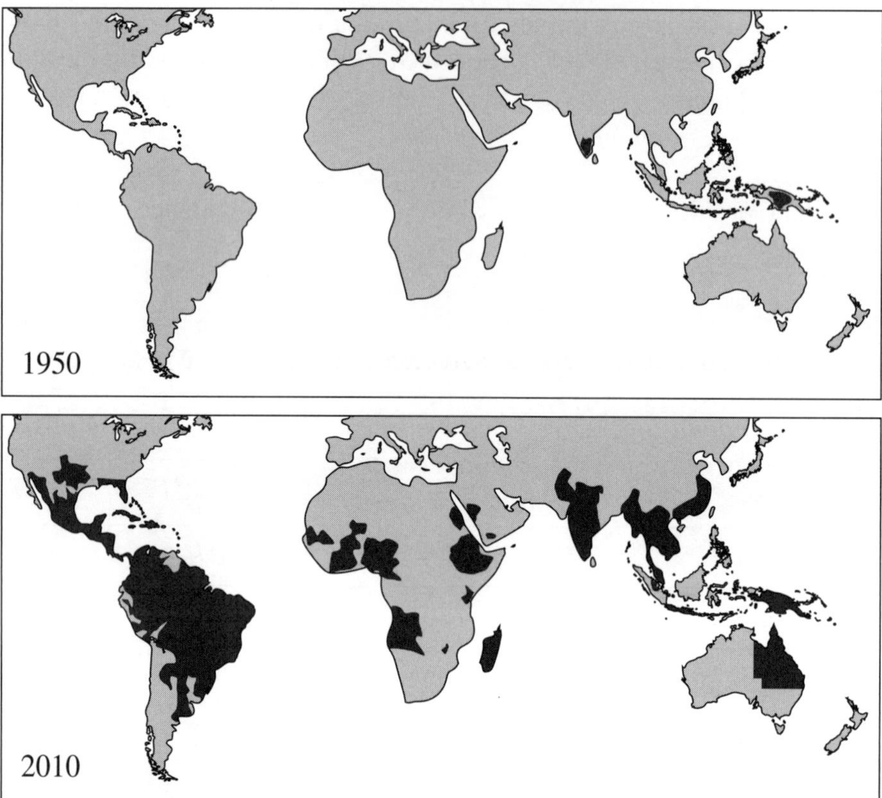

KEY ■ Reported cases of the viral disease dengue fever

(a) Based on the data provided, identify trends in the global disease burden for both malaria and dengue fever. **3**

....

....

....

....

....

....

....

Question 32 continues

Question 32 (continued)

(b) Analyse factors that could have contributed to the change in global distribution of both dengue fever and malaria over the last 100 years. Support your answer with reference to the data provided. **7**

Question 32 (continued)

End of Question 32

Please turn over

Question 33 (20 marks)

Alzheimer's disease causes destruction of brain tissue, dementia and eventually death.

The diagram shows the effect of Alzheimer's disease on the brain.

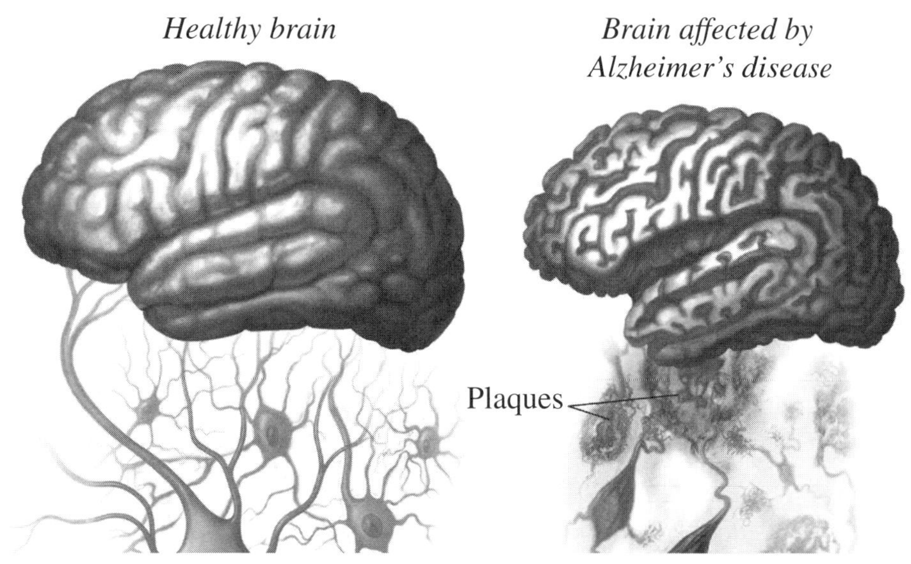

(a) Amyloid beta protein is produced in the human brain throughout life. In people with Alzheimer's disease, it accumulates in excessive amounts. **3**

Outline the main steps that brain cells use to make proteins such as amyloid beta.

....

....

....

....

....

Question 33 continues

Question 33 (continued)

(b) The gene with the greatest known effect on the risk of developing late-onset Alzheimer's disease is called APOE. It is found on chromosome 19.

The APOE gene has multiple alleles, including e2, e3 and e4.

(i) What are multiple alleles? **2**

....

....

....

....

(ii) The table shows the risk of developing Alzheimer's disease for various APOE genotypes compared to average risk in the population. **4**

APOE genotype	e2/e2	e2/e3	e2/e4	e3/e3	e3/e4	e4/e4
Risk of developing Alzheimer's disease (compared to average)	40% less likely	40% less likely	2.6 times more likely	Average	3.2 times more likely	14.9 times more likely

Analyse the data to assess the risk of developing Alzheimer's disease associated with the e2, e3 and e4 alleles.

....

....

....

....

....

....

....

....

....

Question 33 continues

Question 33 (continued)

(c) A large epidemiological study was conducted. It used historical data to investigate the association between *Herpes simplex* virus (HSV) infection and dementia. Dementia is caused by a variety of brain illnesses. Alzheimer's disease is the most common cause of dementia. **3**

The study used the records of 8362 patients with HSV infection and 25 086 randomly selected sex- and age-matched control patients without HSV infection. Some of the patients with HSV had been treated with antiviral medication.

The graph below shows some results of the study.

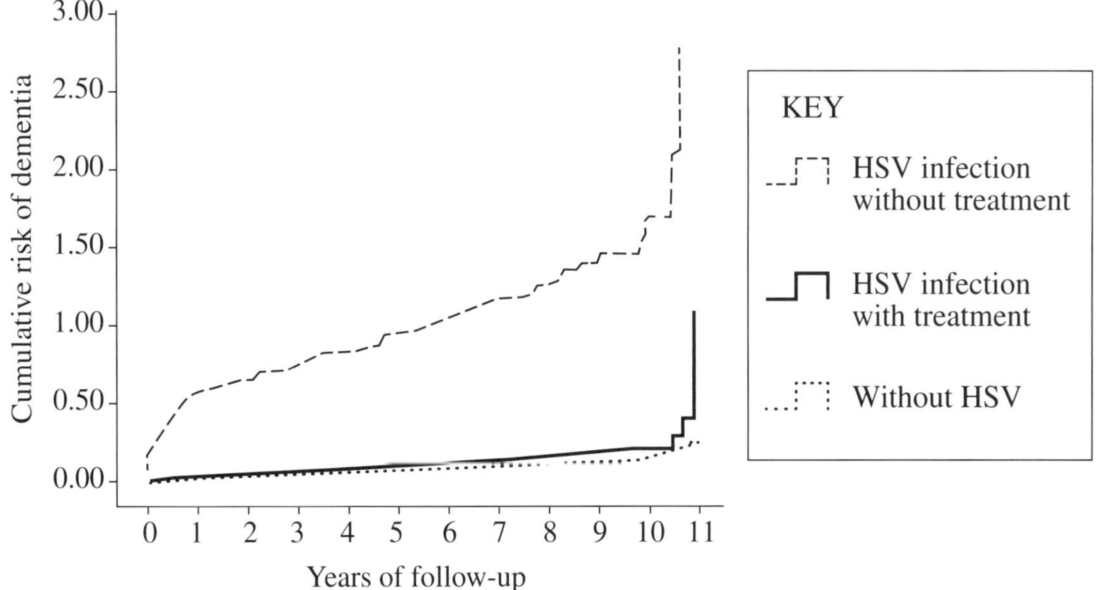

Describe the trends shown in the data.

....

....

....

....

....

....

....

....

....

Question 33 continues

Question 33 (continued)

(d) Diseases are classified as infectious or non-infectious. **8**

Evaluate whether Alzheimer's disease should be classified as an infectious disease or a non-infectious disease. In your answer, include reference to the information and data provided throughout Question 33.

End of paper

2019 HSC Examination Paper

Sample Answers

Section I (Total 20 marks)

1 D Gene mutations contribute to inherited (not infectious) diseases.

2 C Antibodies are produced in response to the vaccine that contains an antigen.

3 D Birds' feeding provides environmental pressure that acts through natural selection.

4 D A nucleotide is the monomer that makes up the DNA polymer. It consists of a sugar (pentagon), a phosphate group (circle) and a nitrogenous base (rectangle or square).

5 C The innate immune response is non-specific. Stomach acid acts indiscriminately on pathogens.

6 C Cochlear implants virtually bypass the ear to stimulate the auditory nerve.

7 A Antibiotic *W* was moderately effective in retarding the growth of both bacteria *A* and bacteria *B* as evidenced by the relatively large area surrounding the antibiotic discs (darker area) with no bacterial growth.

8 B This is the most appropriate type of graph to demonstrate the various numbers of species and reflect biodiversity (as opposed to numbers of animals that reflect population size).

9 A Internal fertilisation reduces the exposure of gametes to the external environment.

10 C The quarantine policy is the independent variable that impacts on the dependent variable, the number of diseased cattle.

11 B The definition of a dominant allele is that it is expressed in the heterozygous genotype.

12 D In type 2 diabetes the blood glucose level will increase rapidly after consuming glucose and it will decrease slowly as the person respires, using some of the blood glucose.

13 B Genetic drift results in changes in allele frequency due to chance or random events, usually in small populations.

14 A The anticodon comprises RNA in which the thymine (T) of the DNA codon is replaced by uracil (U). The 'A' in the codon therefore is coded by 'U' in the anticodon.

15 C Non-coding stretches of DNA may not affect protein structure, although they can affect gene regulation.

16 D Student *X* will respond with a much higher level of antibodies, similar to secondary exposure to the pathogen because of the vaccination. Student *Y* should produce antibodies similar to that of primary exposure in the graph.

17 A The disorder could be autosomal dominant, because if both affected parents have an unaffected offspring it cannot be recessive. If it is sex-linked it must be dominant for both affected parents to produce an unaffected son (the mother must have had one unaffected X chromosome). Also the affected father passed the condition to both daughters.

18 B Until birth has occurred the feedback loop is positive, enhancing the impact of oxytocin on uterine contractions. Negative feedback loops restore homeostasis in processes such as temperature regulation.

19 A Cas9 promotes the cutting of DNA and remains unchanged so its role is that of an enzyme.

20 B CRISP/Cas9 allows more precise gene editing but the Cas9 does the cutting.

Section II

Question 21 (Total 3 marks)

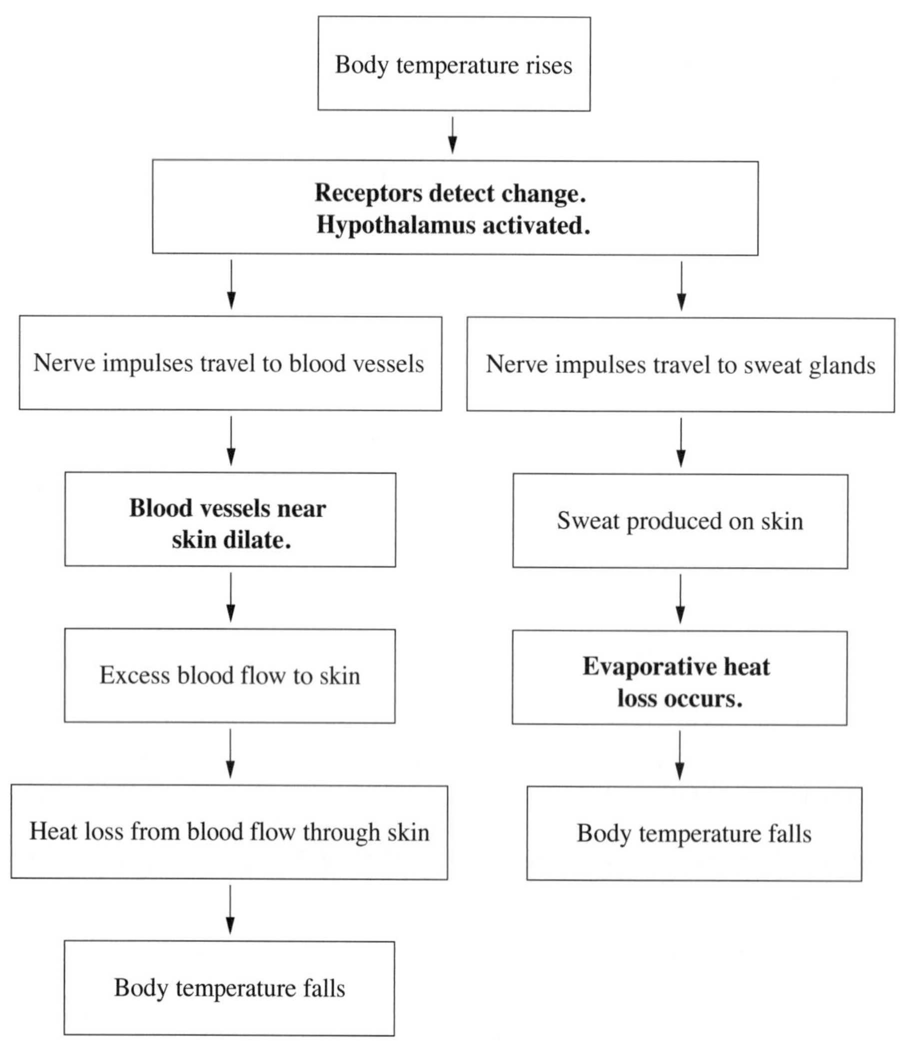

(*3 marks*)

Question 22 (Total 3 marks)

	Somatic mutation	*Germ-line mutation*
Location	Normal body cells	Gametes
Effect on offspring	Nil	Can be significant resulting in inherited disorders as mutation in germ cells can be passed on to offspring and affect all their somatic cells.
Example	Ultraviolet light causes mutations in skin cells resulting in melanoma (skin cancer).	Haemophilia can result from a germ line mutation on the X chromosome. It results in impaired blood clotting.

(*3 marks*)

Question 23 (Total 5 marks)

Educational programs can be effective in reducing the incidence of non-infectious diseases by:

(a) promoting healthy behaviours such as improved diets resulting in less malnutrition and obesity, increased activity levels resulting in improved cardiovascular health, and the use of sun protection resulting in reduced risk of melanoma. Reduced levels of smoking, resulting from ongoing anti-smoking campaigns, has caused a reduced incidence of lung cancer

(b) increased public awareness of risks from environmental exposure resulting in pressure on governments to legislate for occupational health and safety such as banning smoking in public places and implementing laws about safe disposal of asbestos

(c) increased participation in screening resulting in earlier detection and better treatment of cancers such as breast and prostate cancer. Educational programs target female participation in breast screening and recently men in having prostate examinations

(d) increased understanding of inherited disorders and more informed decisions by people at risk of having offspring with an inherited disease. (*5 marks*)

Question 24 (Total 5 marks)

Biodiversity can be reduced because of the reduction of ecosystem diversity (for example, from land clearing), species diversity (for example, from monoculture agriculture) and generic diversity.

Two technologies used in agriculture that can reduce genetic biodiversity are as follows:

(1) Selective breeding reduces biodiversity by reducing some allele frequencies in the gene pool while increasing other allele frequencies that are linked to more economically desirable agricultural products. Selective breeding has in recent times had a greater impact on biodiversity reduction through the use of artificial insemination, IVF, oestrus induction and synchronisation and MOET (multiple ovulation and embryo transfer).

(2) Cloning reduces biodiversity by creating offspring that are genetically identical to a single parent. Vegetative propagation has been used extensively in horticulture to produce large quantities of genetically identical (and hence predictable) plants. Recently many types of agricultural animals have been cloned through the process of somatic cell nuclear transfer. Cloned animals have been used in agriculture for breeding. (*5 marks*)

Question 25 (Total 5 marks)

(a) The evidence of chromosomal mutation in the displayed karyotype is the presence of only one sex chromosome (i.e. aneuploidy). *(1 mark)*

(b) Cell division and fertilisation can lead to the production of this karyotype by non-disjunction of sex chromosomes during meiosis. Gametes resulting from non-disjunction would have either a missing or additional sex chromosome. Non-disjunction could occur in stage one anaphase when homologous pairs segregate or stage two anaphase when 'sister' chromatids separate. The karyotype displayed would result from fertilisation of the gamete containing one fewer chromosome with a normal gamete. *(4 marks)*

Question 26 (Total 5 marks)

The map shows that the indigenous populations in north-west Europe, pockets of the Middle East and central Africa have largely retained the ability to digest lactose with a transition to reduced ability to digest lactose the further away from these centres. The failure of the lactase gene to be switched off in the population of north-west Europe could result from a mutation in the non-coding part of DNA that controls the expression of the lactase gene. Evolution helps to explain the distribution shown on the map because it could be assumed that there was a selective advantage of this mutation in north-west Europe, pockets of the Middle East and central Africa where there was increased dependence on dairy foods compared with other regions. These patterns reflect changes in allele frequency that are evidence of human migration and gene flow. *(5 marks)*

Question 27 (Total 5 marks)

(a) *Budding* is a type of asexual reproduction in which a cell or organism divides asymmetrically into two cells or organisms of unequal size. *(1 mark)*

(b) To test the effect of temperature on reproduction in yeast you would design a controlled investigation in which temperature is the independent variable and the population size of yeast the dependent variable. All other conditions for the yeast such as the type and initial size of the yeast population, moisture, food availability, environmental exposure of humidity, light and air movement would be controlled (kept the same).

An important factor in the investigation would be the method used to measure reproduction in yeast. Microscopic counting of samples of the yeast medium is one potential method. If the yeast released carbon dioxide then the rate of carbon dioxide release could be a relative measure of the size of yeast population and indirectly yeast reproduction. *(4 marks)*

Question 28 (Total 6 marks)

(a)

Alleles from father	*Alleles from mother*
H R	h r

(2 marks)

(b)

		KEY
Homologous pair of chromosome 4 before crossing over	H H R R h h r r	result of crossing over
Homologous chromosomes after crossing over and separation	H H R r h h R r	result with no crossing over
Gametes	H R H r h R h r	

(4 marks)

Question 29 (Total 3 marks)

One mechanism by which plants maintain internal water homeostasis is through the opening and closing of stomata as a result of changes to the shape of guard cells. When water is available the guard cells are turgid and their pressure allows the cell wall closest to the stomata to curve, opening the gap. When moisture level is low the flaccid guard cells close the stomatal opening. Some of these plants open stomata at night when it is cooler. They have the ability to carry out the steps of photosynthesis that use carbon dioxide at night by storing products from the light reaction. *(3 marks)*

Question 30 (Total 5 marks)

The crossed F1 seed shape plants achieved a 3:1 ratio, typical of a cross between F1 parent heterozygous for a dominant allele. Say round is dominant (R) and wrinkled (r) recessive for seed shape, then the Punnett square that explains the result is:

	R	r
R	RR	Rr
r	Rr	rr

As round is dominant, the phenotypic ratio of 3:1 is explained (3-RR, Rr, Rr : 1 rr). The phenotype of the heterozygous genotype is the same as the phenotype of the homozygous dominant allele.

The feather colour F2 generation exhibited a 1:2:1 phenotypic ratio. Assume R is red, W is white.

	R	W
R	RR	RW
W	RW	WW

The phenotype ratio of 1:2:1 resulting from the cross of two heterozygous parents could be explained if the alleles are co-dominant, i.e. both are expressed in the heterozygote so that they have a phenotype that is different from either homozygous phenotype.

The ratio is 1 RR – red : 2 RW – 'roan' : 1 WW – white. The 'roan' chickens would have a mixture of some white feathers and some red feathers. If co-dominance accurately explains the 'roan' phenotype, both heterozygous parents would have a 'roan' phenotype. Incomplete dominance is a similar pattern in which the resultant heterozygous offspring have a 'blended' phenotype of F1 homozygous parents' phenotype. (*5 marks*)

Question 31 (Total 5 marks)

(a) One adaptation of a pathogen, *Plasmodium,* a protozoan that causes malaria and facilitates its entry into a host, is the ability through several forms in its life cycle to survive in the vector mosquito (*Anopheles*) as well as infected humans. The pathogen readily enters the blood when a female mosquito bites a human. (*2 marks*)

(b) The main modes of transmission of pathogens are either direct contact, indirect contact or vector transmission. The mode of transmission of pathogens influences the spread of diseases because direct contact is limited to transmission through touching or droplets from coughing or sneezing within a one-metre range. Spread of disease is limited by the mobility and hygiene behaviour of patients.

Indirect contact can result in disease spreading rapidly over large distances. Public places allow extensive airborne transmission as well as opportunities for transmission on fomites. Contaminated food or water also can result in rapid and widespread transmission.

Vector transmission spreads disease due to the mobility and life cycle of the vector. Fleas, mosquitoes, ticks and mites can carry pathogens across borders. Rats dispersed fleas widely resulting in the widespread transmission of the bubonic plague. (*3 marks*)

Question 32 (Total 10 marks)

(a) Disease trends for malaria between 1900 and 2010 show there is a dramatic reduction in the number of countries reporting an incidence of malaria as well as a reduction of the percentage of the world population at risk of contracting malaria. Dengue fever shows contrasting trends of increased global burden between 1950 and 2010 with a vastly increased distribution of the disease through the tropics. (*3 marks*)

(b) Factors that could have contributed to the reduced distribution of malaria include:

(i) an increased range and effectiveness of methods used to control the spread of the mosquito vector. These include environmental controls, insecticides, bed nets, protective clothing and repellents

(ii) the development of drugs or vaccines that provide resistance in humans to the protozoan pathogen

(iii) the development through biotechnology of genetic changes to either the pathogen or the vector that prevent stages of the life cycle of the pathogen occurring in the vector or the vector's reproduction.

Factors that could have contributed to the increased distribution of dengue fever include:

(i) a reduced effectiveness of methods used to control the spread of the mosquito vector. This could include the mosquito becoming resistant to insecticides. Over an extended period, it is also possible that the mosquito evolved to better adapt to urbanisation or to make use of other potential breeding sites

(ii) changes in environmental conditions, such as increased temperature and rainfall in some areas that increased the potential range of distribution for the mosquito vector

(iii) the increased mobility of carriers who pass the virus back to mosquitoes. This could include forms of the disease that allow patients to live longer or remain asymptomatic and so render quarantine less effective in stopping the spread of the disease. Patients who live longer because of better treatments to manage symptoms (but not kill the virus) may travel and also contribute to the increased distribution of the disease. (*7 marks*)

Question 33 (Total 20 marks)

(a) The main steps that brain cells use to make proteins is via a process called polypeptide synthesis that is described as two sequential processes: transcription followed by translation.

Transcription starts when the DNA in the brain cell nucleus unwinds and the two strands separate. Messenger RNA (mRNA) is a single-stranded nucleotide polymer that forms. This carries the code in the form of codons or triplets of nucleotide bases. Messenger RNA undergoes a process of editing.

Once through the nuclear pores and in the cytoplasm the mRNA attaches to a subunit of ribosome so that *translation* commences. Particular units of transfer RNA (tRNA) attach to specific amino acids and they contain anticodons that slot into the codons of mRNA. The amino acids join together to form a polypeptide.

Polypeptide chains become folded into secondary and tertiary structures and may combine with other polypeptide chains and become a complete protein. (*3 marks*)

(b) (i) Multiple alleles occurs when diploid organisms have more than two forms of the allele within the gene pool. Human blood groups have alleles A, B and O that in combination produce a number of blood types (A, B, AB and O). (*2 marks*)

(ii) The allele e3 when homozygous results in an average risk of developing late-onset Alzheimer's disease. The allele e2 when homozygous results in a greatly (40%) reduced risk, as also when occurring in heterozygous form with the e3 allele. The e4 allele when occurring as homozygous greatly (about 15 times) increases the risk of developing Alzheimer's disease. When it occurs in the heterozygous genotype with the e2 allele the increased risk is only 2.6 times and when with e3 the increased risk compared with average is 3.2.

The e4 allele appears to have a great impact resulting in increased risk of late-onset Alzheimer's disease. Allele e2 has the greatest impact in reducing the risk of late-onset Alzheimer's disease. (*4 marks*)

(c) The trends shown in the graph indicate that patients with untreated HSV infection are at greatest risk of dementia. Patients with treated HSV infections maintain a relatively low risk for 10 years. Both treated and untreated HSV patients find a sharp cumulative increased risk of dementia after 10 years of follow-up. (*3 marks*)

(d) Non-infectious diseases such as inherited diseases are not transmitted by pathogens nor are they transmitted through contact or vectors. Evidence exists that patients have an increased risk of late-onset Alzheimer's disease (one form of dementia) in the presence of allele e4 for gene APOE on chromosome 19. It could be argued that many patients suffering dementia as a result of late-onset Alzheimer's disease are experiencing a non-infectious inherited disease. More evidence that would support this hypothesis could come from establishing biochemical pathways between the action of the e4 allele of APOE and the excessive production of amyloid beta protein in brain cells.

Evidence that after 11 years of study patients who had experienced HSV (*Herpes simplex* virus) either treated or untreated showed a dramatically increased risk of dementia suggests a correlation between both HSV and dementia. This correlation is further strengthened by analysing data for patients of untreated HSV, who experience an increased risk of dementia throughout the entire period of the follow-up data collection. The fact that both graphs connecting HSV infection and dementia showed dramatic increases after 11 years would suggest that the form of dementia experienced could have been late-onset Alzheimer's disease.

Consideration needs to be taken into account that the epidemiological study does not suggest a direct cause and effect but only finds a correlation between factors. Further investigation is needed to establish whether the HSV brings about changes in the brain to result in Alzheimer's. Another limitation of this evidence is the lack of data as to the proportion of dementia patients studied that experienced the late-onset Alzheimer's form of dementia.

There is increasing evidence from diseases such as cervical cancer that are linked to HPV (human papilloma virus) infection that infectious viral diseases can result in mutations that cause cancer.

Is late-onset Alzheimer's non-infectious? The evidence demonstrates a highly probable inherited cause in some Alzheimer's patients, i.e. that it is non-infectious. More evidence needs to be gathered about other possible environmental conditions that might contribute to the onset of Alzheimer's disease that would strengthen the case for it being a non-infectious disease. If a biochemical link could be made between the e4 allele for the APOE gene and the production of amyloid beta protein in the brain cells this would support the case for late-onset Alzheimer's being non-infectious.

Evidence that the probability that Alzheimer's disease is infectious would be strengthened in the epidemiological study if there could be identification of the number of dementia patients studied who were sufferers of late-onset Alzheimer's disease. Epidemiological evidence suggests a correlation between suffering an infection (HSV), especially if untreated, and the development of dementia. The conclusion is that the distinction between infectious and non-infectious disease is no longer always accurate. Some diseases such as late-onset Alzheimer's have complex, interacting causes. (*8 marks*)

CHAPTER 10

NSW Education Standards Authority

2020 HIGHER SCHOOL CERTIFICATE EXAMINATION

Biology

General Instructions

- Reading time – 5 minutes
- Working time – 3 hours
- Write using black pen
- Draw diagrams using pencil
- Calculators approved by NESA may be used

Total marks: 100

Section I – 20 marks

- Attempt Questions 1–20
- Allow about 35 minutes for this section

Section II – 80 marks

- Attempt Questions 21–32
- Allow about 2 hours and 25 minutes for this section

20 marks
Attempt Questions 1–20
Allow about 35 minutes for this section

Use the multiple-choice answer sheet for Questions 1–20.

---

1 In maintaining homeostasis, which of the following is a behavioural adaptation?

A. Sweating to cool down

B. Curling up in a ball to keep warm

C. Speeding up or slowing down cell metabolism

D. Skin going red as more blood flows to surface

2 Sexual reproduction in plants involves

A. pollination caused by dispersal of seeds.

B. cloning as it creates copies of the parent plant.

C. mitosis leading to the formation of pollen grains.

D. fertilisation as a result of fusion of male and female gametes.

3 The following four events occur during reproduction in a placental mammal.

1. Fertilisation
2. Implantation
3. Ovulation
4. Placental formation

In which order do these events occur?

A. 2, 1, 3, 4

B. 2, 4, 1, 3

C. 3, 1, 2, 4

D. 3, 2, 4, 1

4 Malaria is a disease in humans caused by a single-celled *Plasmodium* species. It is transmitted by female mosquitoes.

Which of the following is true for malaria?

A. Both *Plasmodium* and the mosquito are vectors

B. Both *Plasmodium* and the mosquito are pathogens

C. The mosquito is the vector and *Plasmodium* is the pathogen

D. The mosquito is the pathogen and *Plasmodium* is the vector

5 Which row of the table best describes DNA in both prokaryotic and eukaryotic cells?

	Prokaryotic	*Eukaryotic*
A.	Circular	Circular
B.	Circular	Linear
C.	Linear	Circular
D.	Linear	Linear

6 Citrus canker is a bacterial disease that originates in south-east Asia and affects citrus fruit.

What would be the most effective way to prevent the disease from spreading into or across Australia?

A. Monitor citrus trees and fruit continuously.

B. Certify orchards before fruit is transported.

C. Keep citrus trees and fruit entering Australia in quarantine stations until the incubation period has passed.

D. Inspect citrus trees and fruit entering Australia in quarantine stations before transportation across Australia.

7 Students designed and conducted an investigation to test for the presence of microbes in THREE different food samples.

They inoculated agar plates with the samples and placed them in an incubator set to 25°C.

Which row of the table represents a valid design for the investigation?

	Independent variable	*Dependent variable*	*Experimental control*
A.	Food sample	Number of microbes	An agar plate without a sample
B.	Number of microbes	Food sample	Temperature set to 25°C
C.	Food sample	Number of microbes	Temperature set to 25°C
D.	Number of microbes	Food sample	An agar plate without a sample

8 Quarantine is ineffective as a measure to control non-infectious diseases because they

A. cannot develop in isolation.

B. depend on long-term exposure to a pathogen.

C. may be inherited and affect the organism all their life.

D. may only be treated by genetic engineering altering cells.

9 A public education campaign was developed with the aim of lowering the incidence of skin cancer in the population.

The campaign was adopted Australia wide and is illustrated in the poster.

Which is the best method to measure the effectiveness of the campaign?

A. By measuring exposure to the sun and skin cancer incidence

B. By surveying beachgoers, asking if they remember the campaign

C. By comparing skin cancer incidence before and after the campaign

D. By counting the number of people on the beach wearing hats and sunglasses

10 A farmer intends to artificially inseminate cows with semen from a bull which has been chosen based on characteristics of colour and muscle mass.

The farmer does not know that the bull is heterozygous for a rare recessive allele not previously present in the farmer's cow population.

The introduction of this recessive allele to the population of cows is an example of

A. gene flow.

B. genetic drift.

C. natural selection.

D. selective breeding.

11 The diagram shows a model of the human eye.

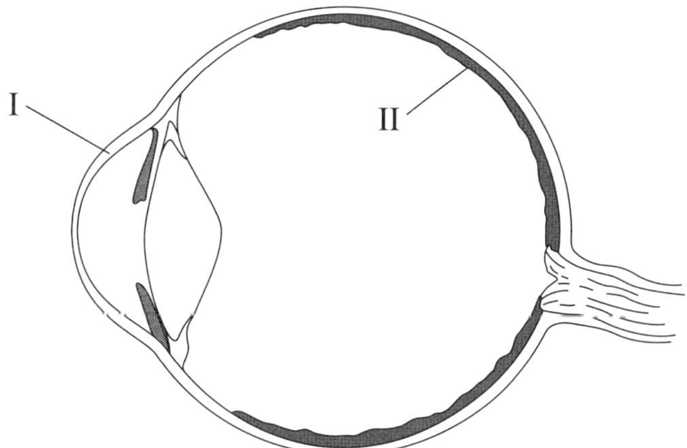

Which of the following correctly identifies a labelled part and its function?

	Label	*Name*	*Function*
A.	I	Cornea	Refract light
B.	I	Retina	Transmit light
C.	II	Retina	Focus light
D.	II	Cornea	Absorb light

12 What is the purpose of cloning in agriculture?

A. Increasing the frequency of recessive traits

B. Preserving favourable traits in the offspring

C. Preserving genetic variability in a population

D. Increasing combinations of alleles in a population

13 A type of genetic technology is shown in the diagram.

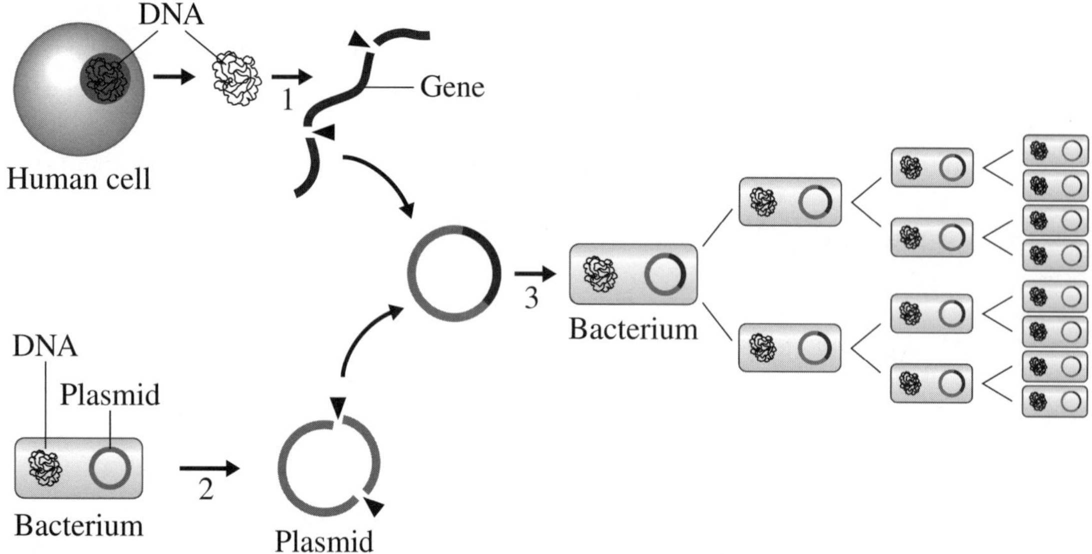

What type of cloning is modelled?

A. Gene cloning because bacteria are used.

B. Gene cloning because a human gene is being replicated.

C. Whole organism cloning because identical offspring are produced.

D. Whole organism cloning because the bacteria use asexual reproduction.

14 A normal allele results in liver cells with sufficient cholesterol receptors. A different allele results in liver cells without cholesterol receptors. Individuals who are heterozygous have liver cells with insufficient cholesterol receptors.

What type of inheritance is the most likely explanation for this?

A. Sex-linked

B. Autosomal dominant

C. Autosomal recessive

D. Incomplete dominance

15 Four antiviral drugs have been tested in a culture of human cells for their effectiveness in inhibiting infection from a new virus. The toxicity of the antivirals to human cells was also tested.

The ability of the drugs to inhibit viral entry to the cells (% inhibition) and the proportion of the cells killed by the drugs (% toxicity) were recorded at different doses of each drug and shown in the graphs.

From the results shown, which antiviral drug is the safest and most effective at a dose of 1 μmol L^{-1}?

A.

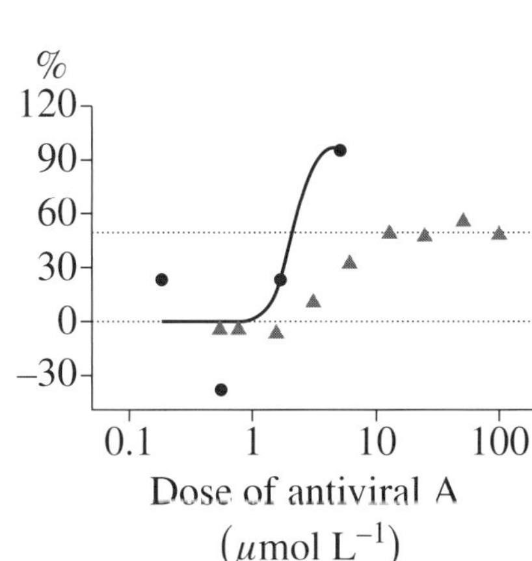

B.

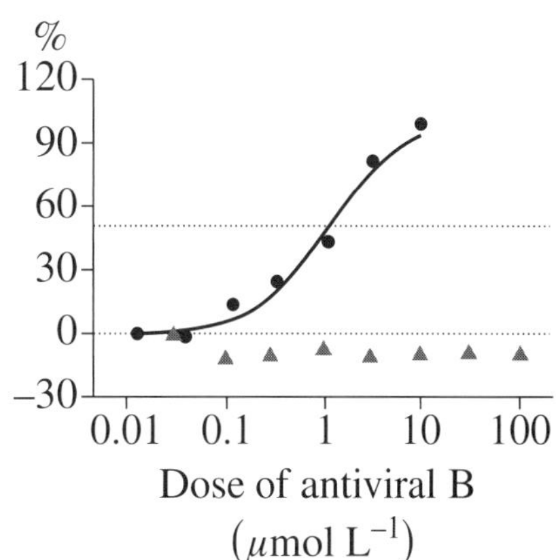

C.

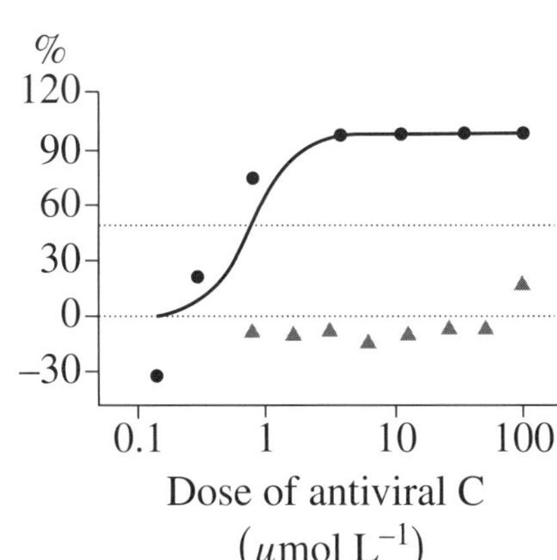

D.

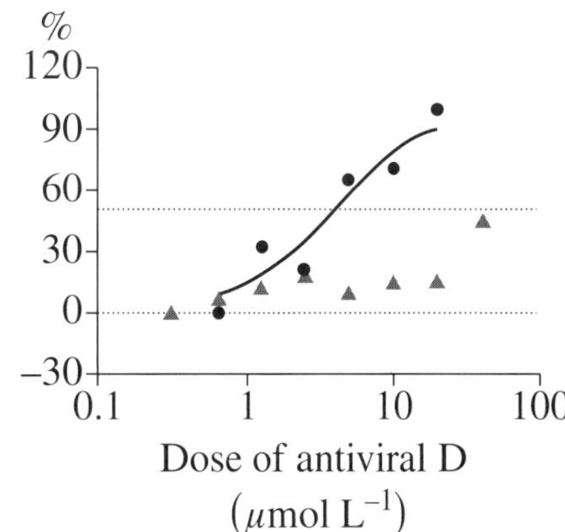

16 Analysis of DNA shows that adenine and guanine always make up 50% of the total amount of nitrogenous bases in DNA.

Which structural feature of DNA does this provide evidence for?

A. DNA is helical in structure.

B. DNA is always a double-stranded molecule.

C. DNA always has adenine paired with guanine.

D. DNA is made up of equal amounts of nitrogenous bases.

17 There are about 10 million single nucleotide polymorphisms (SNPs) found in the human genome.

Four SNPs are modelled in the diagram.

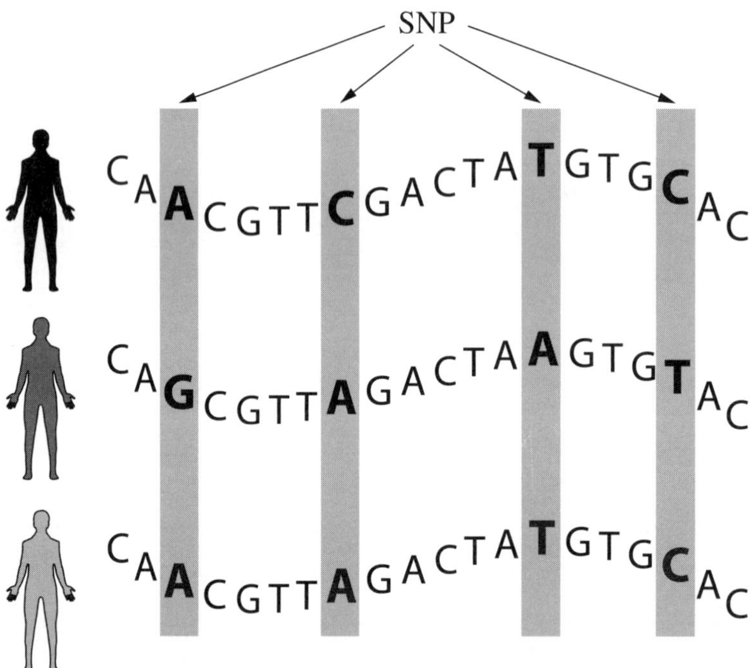

The SNPs modelled do not affect the phenotype of the individuals shown.

Which is the best explanation for this?

A. Only one nucleotide is different at each SNP.

B. The SNPs are part of DNA that is not expressed.

C. AGA, CAA, TAT and CTC all code for the same amino acid.

D. The SNPs are present on one strand of the DNA molecule only.

18 SNP databases have been used in forensic investigations. One is outlined below.

1. DNA was collected at a crime scene 30 years ago.
2. Recently the crime scene DNA was analysed at 700 000 SNP locations.
3. An SNP profile was created and uploaded to a genealogy database.
4. The SNP profile from the crime scene indicated some shared SNPs with two individuals (who did not have SNPs in common).
5. The pedigrees were constructed for the two individuals.

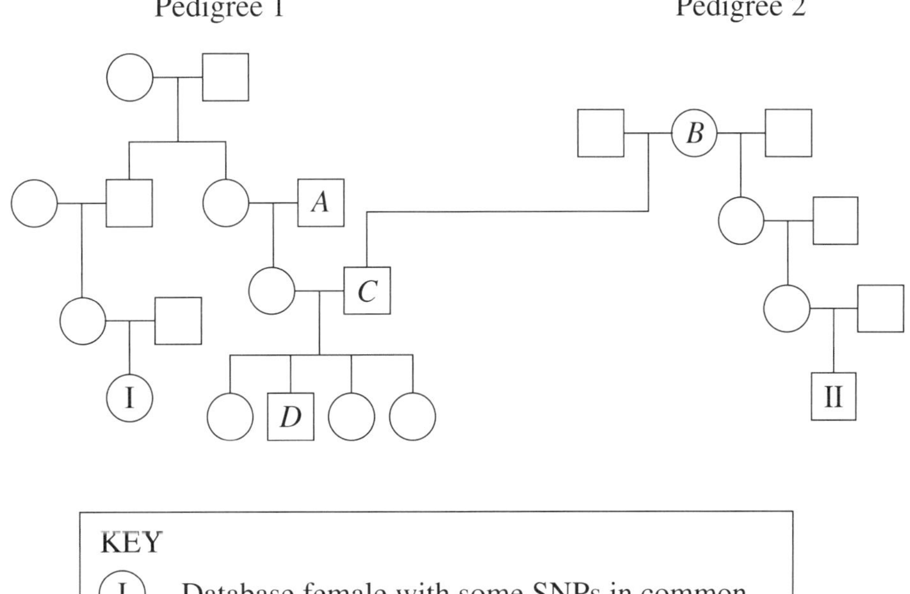

KEY

(I) Database female with some SNPs in common with crime scene DNA

[II] Database male with some SNPs in common with crime scene DNA

Which person is most likely to be the suspect who should be investigated?

A. *A*

B. *B*

C. *C*

D. *D*

19 Which diagram correctly models one phase of meiosis in an organism that has six chromosomes in its somatic cells?

A.

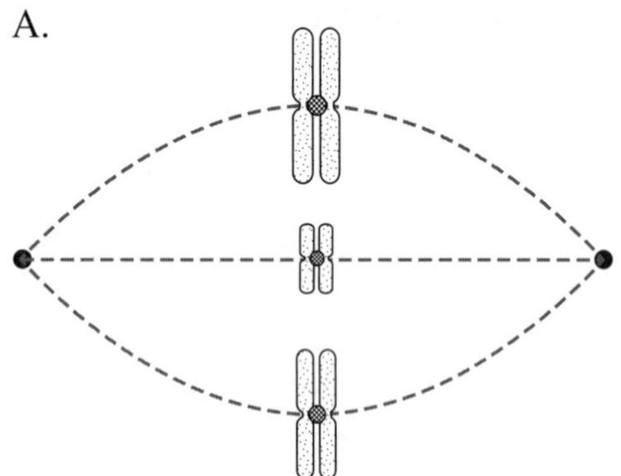

B.

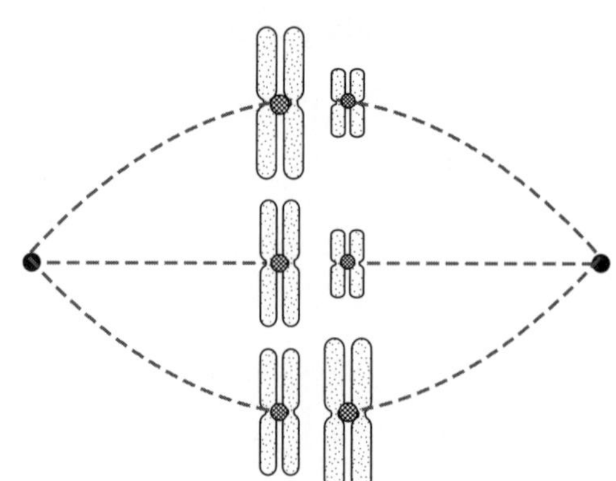

C.

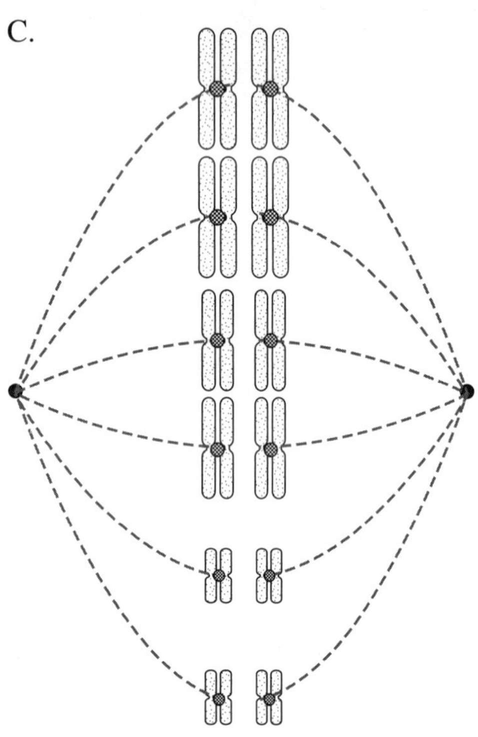

D.

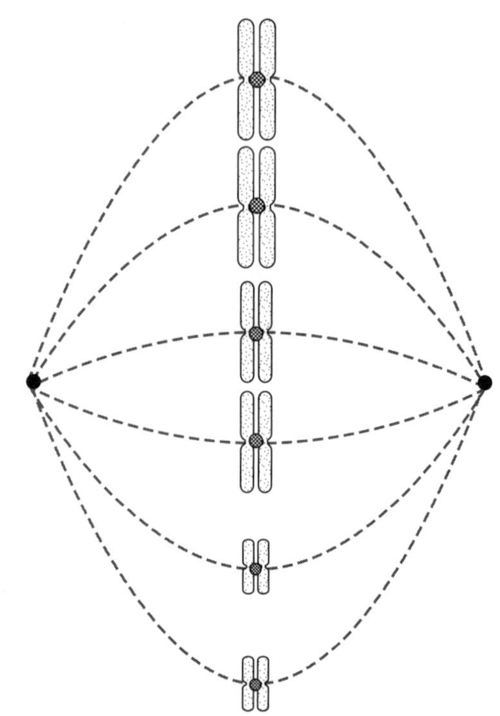

20 This chart illustrates three correlation patterns indicating the influence of genes and environment on different traits in individuals.

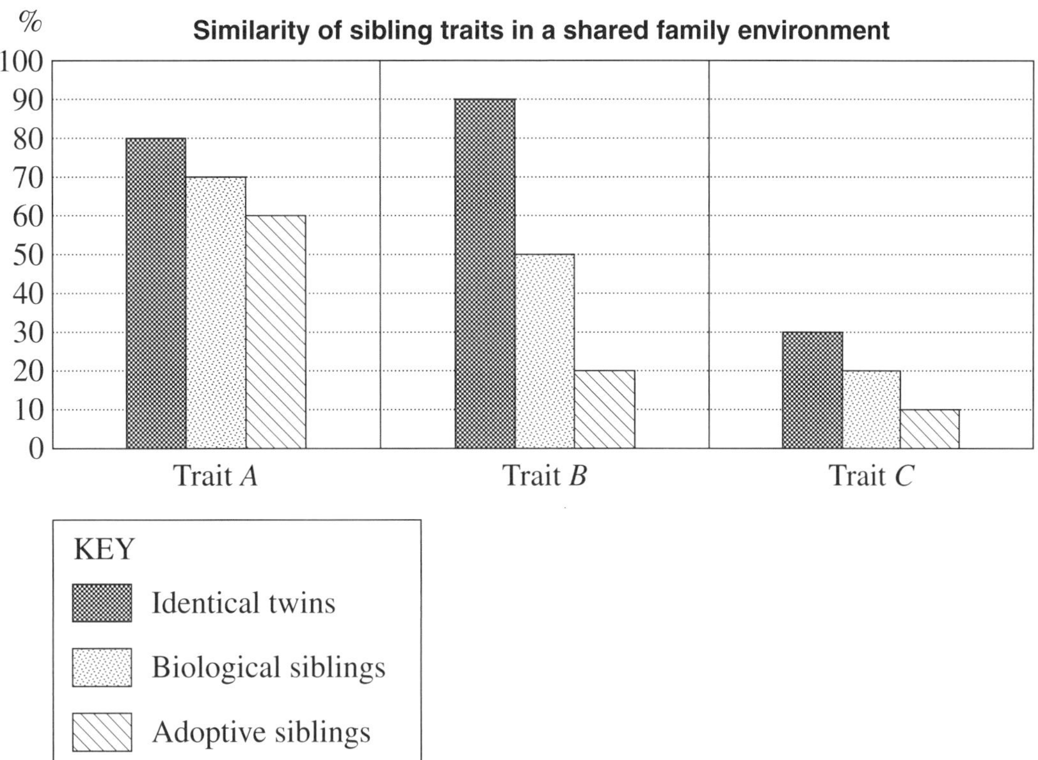

What does the data show about how genes and family environment affect the three traits?

	Trait A		*Trait B*		*Trait C*	
	Genes	*Family environment*	*Genes*	*Family environment*	*Genes*	*Family environment*
A.	Low	High	High	Low	Low	Low
B.	Low	High	High	Low	High	High
C.	High	Low	Low	High	Low	Low
D.	High	Low	Low	High	High	High

2020 HIGHER SCHOOL CERTIFICATE EXAMINATION

Centre Number

Biology

Student Number

Section II Answer Booklet

80 marks
Attempt Questions 21–32
Allow about 2 hours and 25 minutes for this section

Instructions

- Write your Centre Number and Student Number at the top of this page.
- Answer the questions in the spaces provided. These spaces provide guidance for the expected length of response.
- Show all relevant working in questions involving calculations.

Please turn over

Question 21 (3 marks)

Cholera is an acute diarrhoeal infection caused by the bacterium *Vibrio cholerae*. Humans are infected when they consume food or water that is contaminated with the bacterium. **3**

Outline THREE strategies that could prevent the spread of cholera.

....

....

....

....

....

Question 22 (3 marks)

Outline a benefit and a limitation of using pharmaceuticals such as antibiotics to treat infectious disease. **3**

....

....

....

....

....

....

Question 23 (3 marks)

The following diagram shows a mutation.

Original DNA sequence

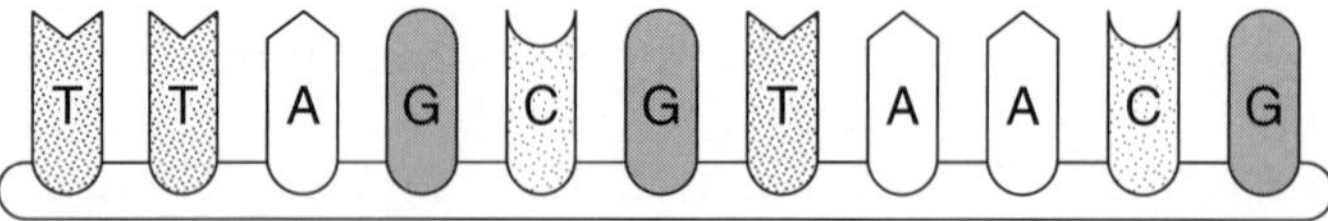

Mutated DNA sequence

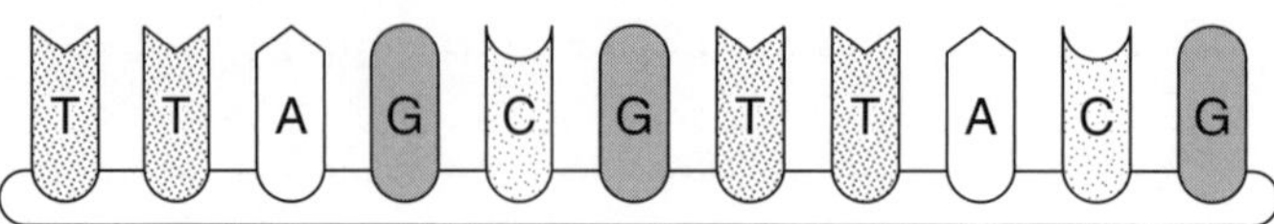

(a) What type of mutation is shown in the diagram? **1**

....

(b) Outline another type of mutation. **2**

....

....

....

Question 24 (7 marks)

An indicator of kidney function is the volume of filtrate formed at the glomerulus in 1 minute (GFR).

GFR of healthy adult	> 100 mL min^{-1}
GFR needing dialysis	< 15 mL min^{-1}

A patient's kidney function was monitored and the following data recorded.

Year	2011	2012	2013	2014	2015	2016	2017	2018	2019
GFR (mL min^{-1})	81	76	77	77	79	65	60	45	35

(a) Plot the data on the grid. **2**

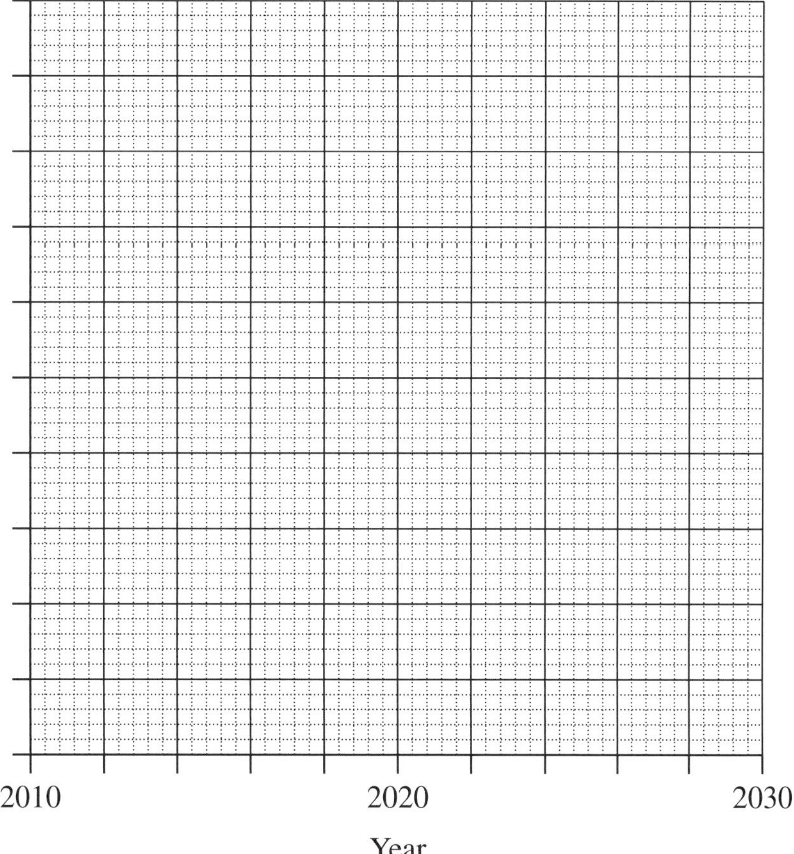

(b) Use the graph to show the year that the patient is predicted to require dialysis. Show your working and answer on the graph. **2**

Question 24 continues on the following page

Question 24 (continued)

(c) Explain how dialysis compensates for the loss of a function of the kidneys. **3**

....

....

....

....

....

....

....

....

End of Question 24

Please turn over

Question 25 (7 marks)

Students tested the hypothesis that the number of eggs/young produced was greater in animals using external fertilisation than those using internal fertilisation. They obtained the following data from secondary sources.

Mode of fertilisation	*Species*	*Average number of young born or eggs laid in one reproductive cycle*	Mean ± SD*
Internal fertilisation	Red kangaroo	1	43 ± 55
	Bush rat	6	
	White tipped reef shark	6	
	Loggerhead turtle	126	
	Red bellied black snake	18	
	Guppy (fish)	100	
External fertilisation	Pouched frog	13	40 ± 32
	Loveridge's frog	20	
	Corroboree frog	25	
	Turtle frog	50	
	Clownfish	100	
	Siamese fighting fish	30	
*SD is standard deviation which gives a measure of the amount of variation in the data.			

(a) What conclusion can be drawn from the data? Justify your answer. **3**

....

....

....

....

....

....

....

Question 25 continues on the following page

Question 25 (continued)

(b) Justify an improvement to the students' experimental design to test the same hypothesis. **2**

....

....

....

....

....

....

....

(c) Explain ONE advantage for animals of using external fertilisation. **2**

....

....

....

....

End of Question 25

Question 26 (6 marks)

One of the genes involved in determining the colour of a species of fish has two alleles: yellow and orange.

The diagram shows a pedigree chart for the inheritance of colour in the fish.

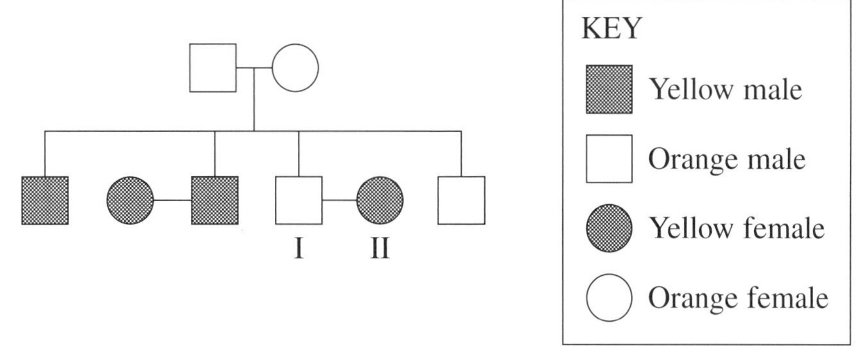

(a) Use the pedigree chart to explain why the yellow allele is recessive. **2**

....

....

....

....

Question 26 continues on the following page

Question 26 (continued)

(b) Explain how a cross between individuals I and II could be used to determine whether the inheritance of colour in the fish is sex-linked or autosomal. **4**

....

....

....

....

....

....

....

....

....

....

....

....

....

....

....

End of Question 26

Question 27 (6 marks)

Exposure to arsenic in drinking water has been associated with the onset of many diseases. The World Health Organisation recommends arsenic levels in drinking water should be below $10\ \mu g\ L^{-1}$.

An epidemiological study involving 58 406 young adults was conducted over an 11-year period in one country to investigate young-adult mortality due to chronic exposure to arsenic in local drinking water. Each individual's average exposure and cumulative exposure to arsenic over the time of the study were calculated. Age, sex, education and socioeconomic status were taken into account during the analysis of the results.

The graphs show survival rates for males and females over the 11-year period associated with different average levels of exposure to arsenic in drinking water.

KEY
- - - - - $< 90\ \mu g\ L^{-1}$ ·········· $90–223\ \mu g\ L^{-1}$ —— $> 223\ \mu g\ L^{-1}$

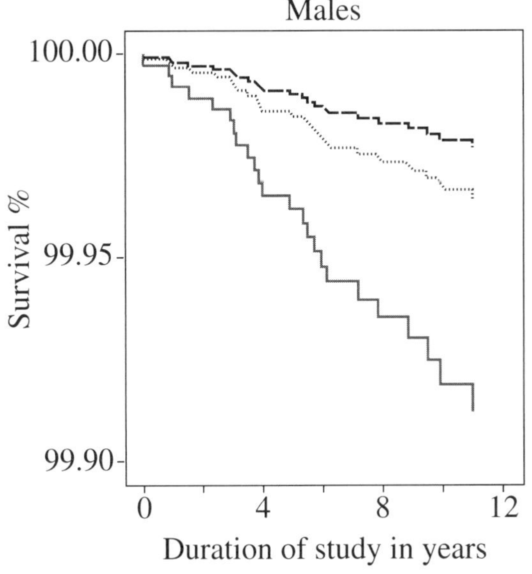

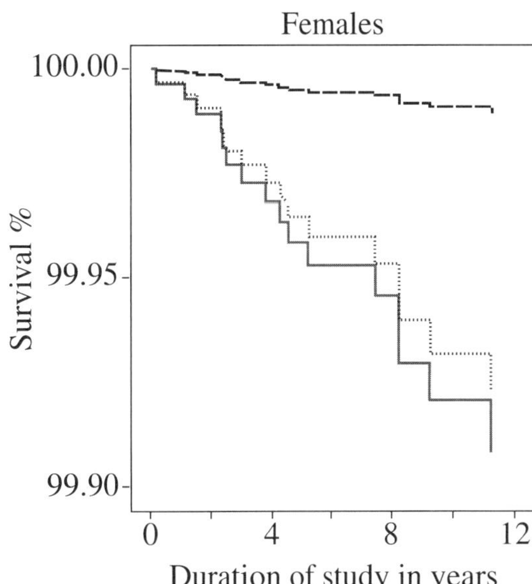

Question 27 continues on the following page

Question 27 (continued)

(a) Identify TWO features of the method used that contributed to the validity of this study. **2**

....

....

(b) The hypothesis put forward was that exposure to arsenic in drinking water increases mortality in young adults. **4**

Discuss the data presented in the graphs in relation to this hypothesis.

....

....

....

....

....

....

....

....

....

....

....

....

....

....

....

....

....

....

....

....

End of Question 27

Question 28 (6 marks)

(a) A student drew a diagram to model part of the process of meiosis. 3

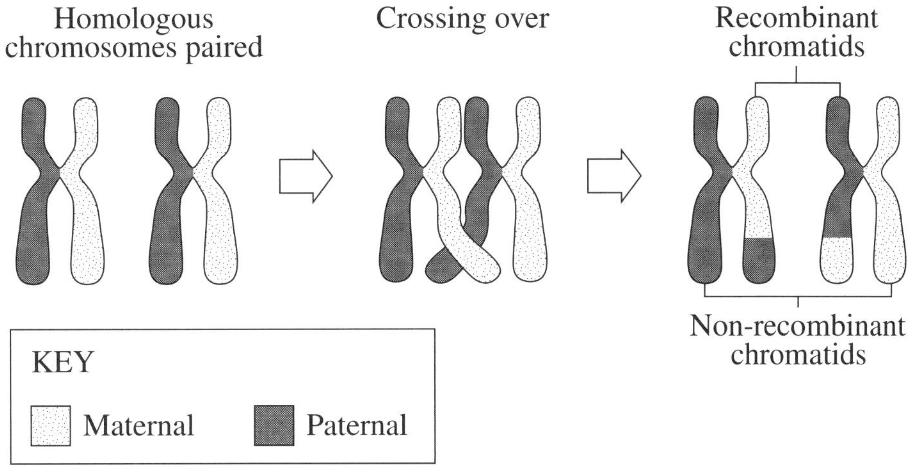

Homologous chromosomes paired

Crossing over

Recombinant chromatids

Non-recombinant chromatids

KEY

Maternal Paternal

Explain the misunderstanding of meiosis shown in this model.

....

....

....

....

....

....

(b) Explain the effect of meiosis on genetic variation. 3

....

....

....

....

....

....

....

....

....

....

Question 29 (5 marks)

Explain how TWO processes that affect the gene pool of populations can lead to evolution. **5**

Question 30 (7 marks)

Explain the impact that genetic technologies have had on the management of both infectious and non-infectious diseases. 7

Question 31 (9 marks)

(a) The levels of glucose, insulin and glucagon were measured in the plasma of 24 healthy adults at intervals over a 5-hour period. After 1 hour at rest the patients ate a large carbohydrate meal. The results are shown. **6**

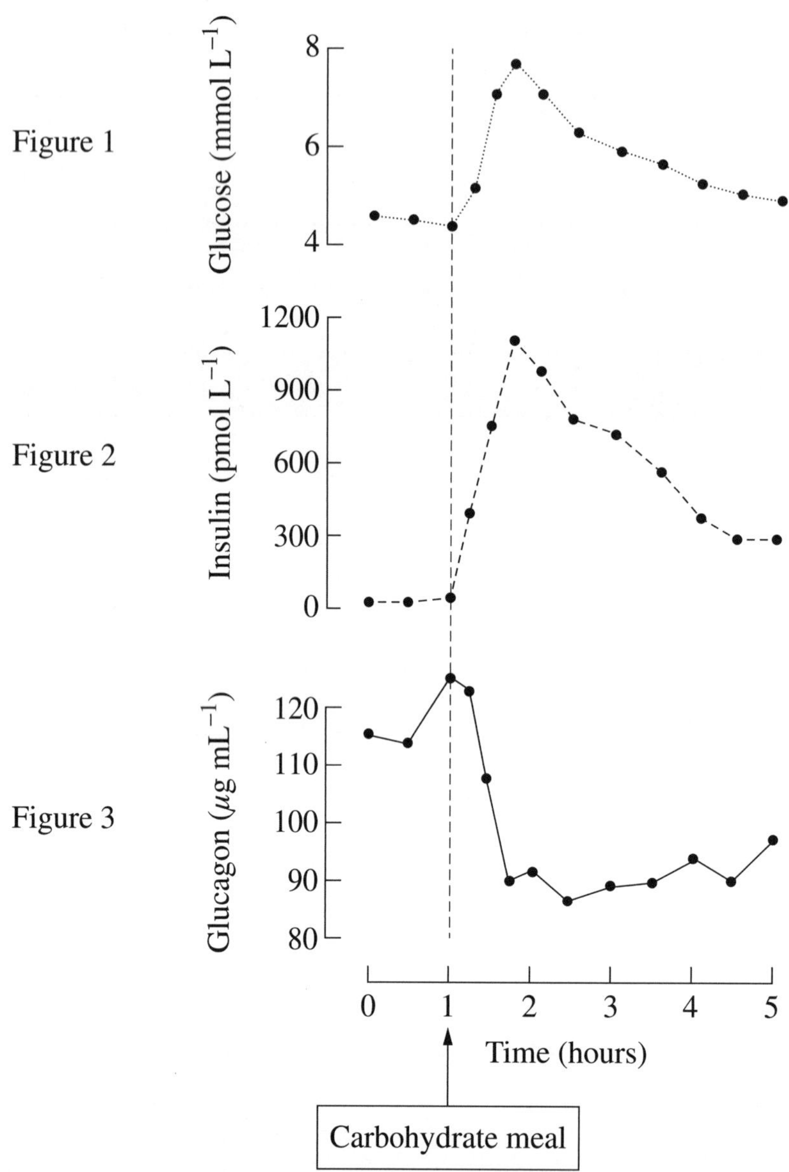

Question 31 continues on the following page

Question 31 (continued)

Use the data provided to explain how blood glucose is controlled in the body.

....

....

....

....

....

....

....

....

....

....

....

....

....

....

....

....

....

....

....

....

....

....

....

....

....

....

Question 31 continues on the following page

Question 31 (continued)

(b) Outline how in humans, maintenance of temperature is different to the way that glucose is controlled. **3**

....

....

....

....

....

....

....

....

End of Question 31

Question 32 (18 marks)

(a) Rabies is a disease that can affect all mammals and is caused by the rabies virus. It is transmitted by the bite of an infected animal. Without treatment it almost always results in death. **2**

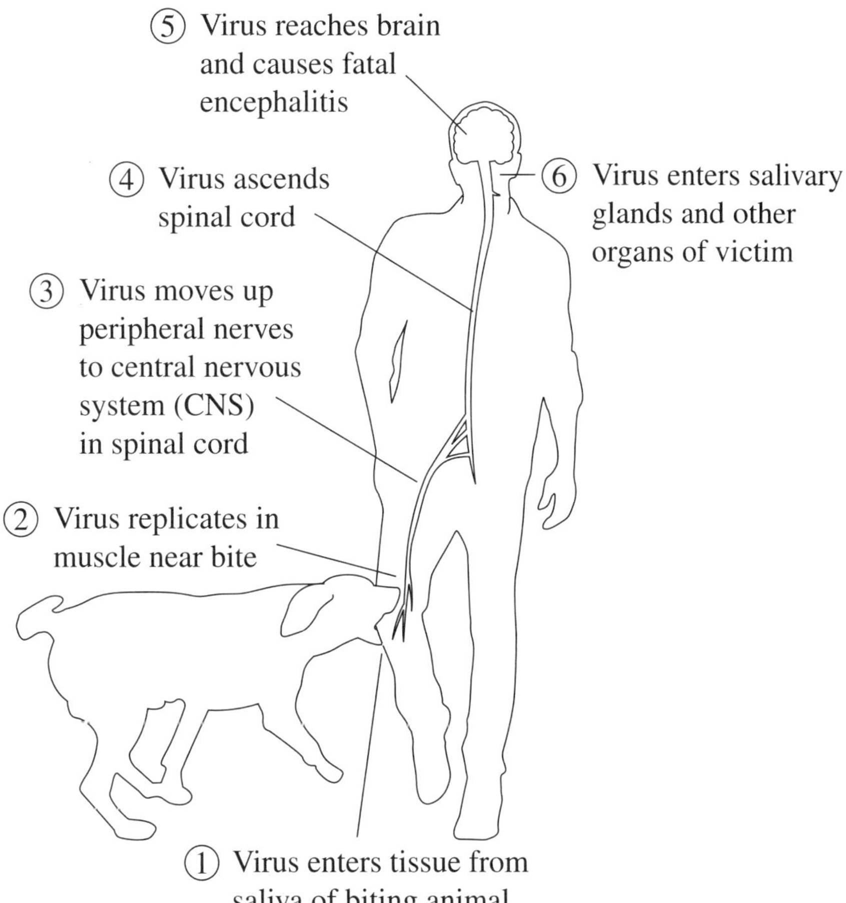

Use the information provided to identify TWO features of the rabies infection that facilitate transmission of the pathogen to a new host.

....

....

....

....

Question 32 continues on the following page

Question 32 (continued)

(b) The rabies virus is a single-stranded RNA virus. It contains and codes for only five proteins. The diagrams show the structure and reproduction of the virus.

Diagram 1 – structure

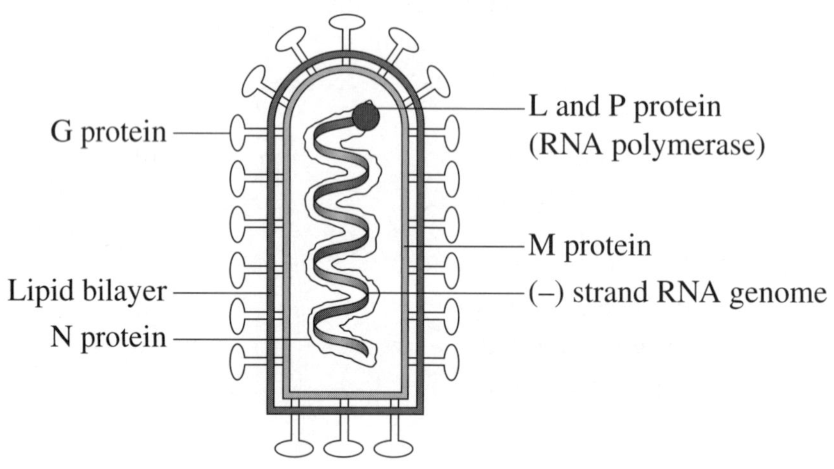

Diagram 2 – reproduction

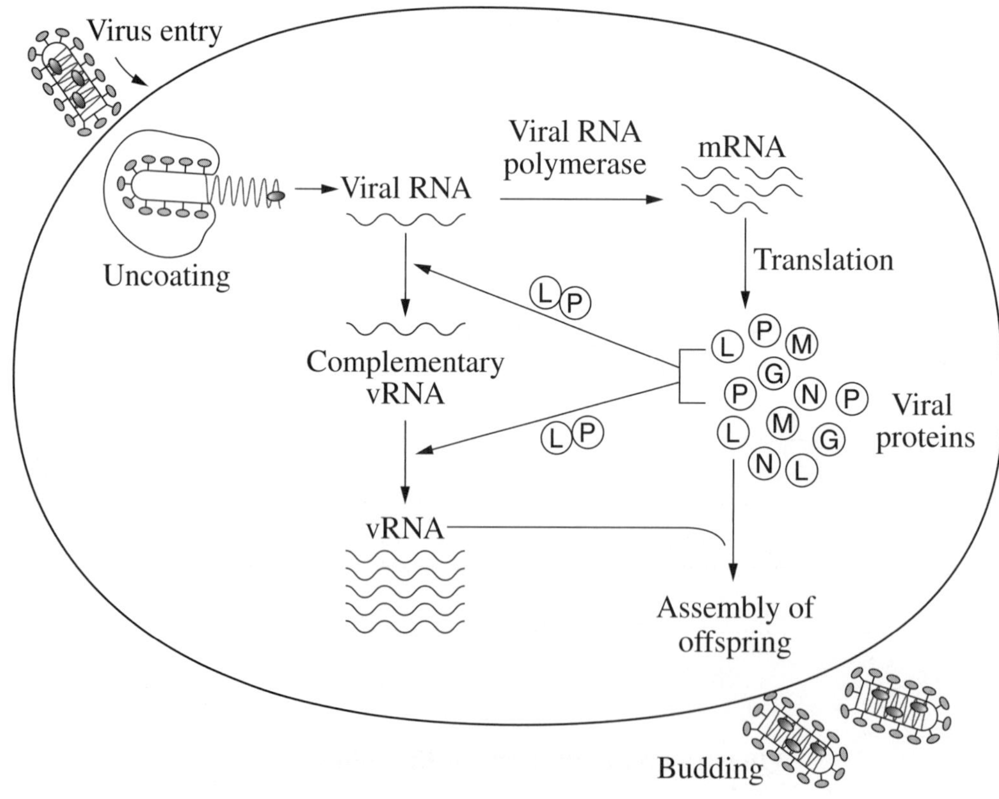

Question 32 continues on the following page

Question 32 (continued)

(i) Use the information provided in Diagram 1 to explain why the rabies virus cannot be classified as a cellular pathogen. **3**

....

....

....

....

....

....

(ii) After infection the virus reproduces in muscle cells near the bite site and in the central nervous system. This requires the single-stranded rabies RNA to be transcribed, translated and replicated in the cytoplasm of host cells. These processes are shown in Diagram 2. **5**

Use the information provided in Diagrams 1 and 2 to explain the role of viral RNA polymerase in the reproduction of the virus.

....

....

....

....

....

....

....

....

....

....

....

....

....

....

....

....

....

Question 32 continues on the following page

Question 32 (continued)

(c) Post exposure prophylaxis (PEP) is given to patients who have been bitten by a rabid animal. **8**

PEP includes an injection of human rabies antibodies (HRIG) as well as injections of a rabies vaccine at 0, 3, 7 and 14 days after exposure to the virus.

The following graphs show a generalised response to rabies infection without and with PEP.

Rabies infection without PEP

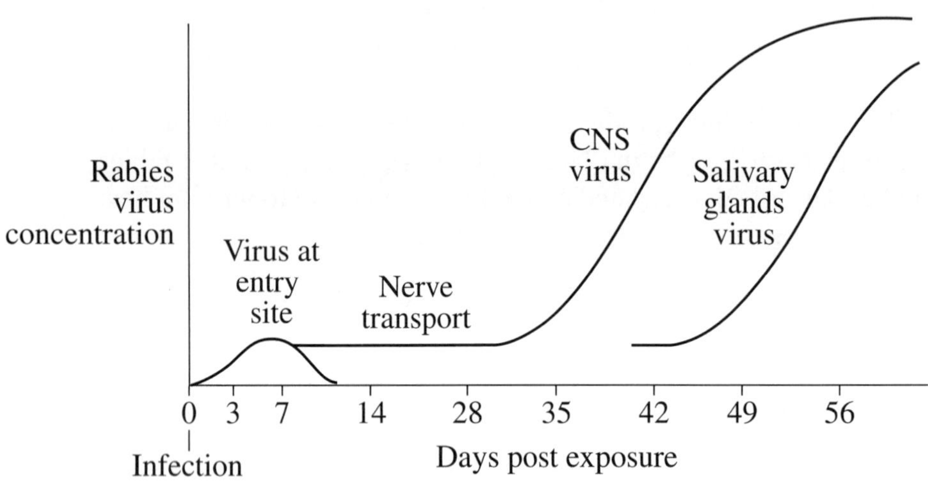

Rabies infection with PEP

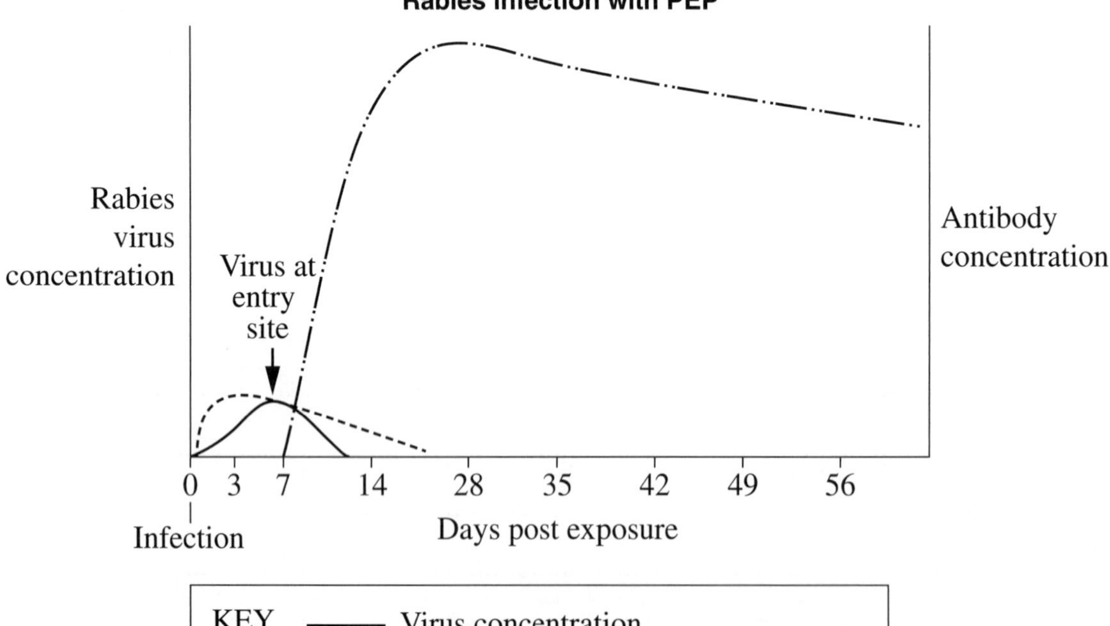

Question 32 continues on the following page

Question 32 (continued)

Explain how PEP prevents rabies developing after infection with the virus. Support your answer with reference to the information and data provided throughout Question 32.

....

....

....

....

....

....

....

....

....

....

....

....

....

....

....

....

....

....

....

....

....

....

....

....

End of paper

2020 HSC Examination Paper

Sample Answers

Section I (Total 20 marks)

1 B Curling up into a ball is a behavioural response of an organism that helps it maintain body temperature.

2 D The key process in sexual reproduction is the union of male and female gametes by the process of fertilisation.

3 C Ovulation in placental mammals must occur first as it results in the production of the female gamete. Once it is fertilised it implants in the uterus wall before the placenta forms.

4 C The *Plasmodium* is the pathogen that when it lives in humans brings about the disease malaria. The mosquito transmits the disease-causing pathogen so it is the vector.

5 B DNA in prokaryotes is mostly contained in a single circular chromosome. Eukaryotes have most of their DNA in densely packed linear chromosomes.

6 C Bacterial infections may not be apparent on initial inspection so quarantining citrus for the duration of the incubation period should enable the detection of disease and prevent it spreading into Australia. If introduced to Australia, continuous monitoring of trees and fruit may assist control.

7 A The food samples are deliberately selected (independent) whereas the number of microbes are the consequence of the contamination in different food samples (dependent). Temperature is a controlled variable. The experimental control, the agar plate without a food sample, is the basis of comparison.

8 C Non-infectious diseases such as inherited diseases are non-communicable and so quarantine is an ineffective control measure.

9 C A reduction in the incidence of skin cancer after the campaign (over an extensive time frame) compared with the incidence before would indicate campaign effectiveness.

10 A Although the introduction of the rare recessive allele was because of the use of artificial insemination it is in effect a form of gene immigration. It is not the result of a random event and was not a consequence of the gene being selected.

11 A The cornea refracts light rays entering the eye.

12 B Cloning is used in agriculture to increase numbers of offspring with favourable traits.

13 B A human gene is cut and pasted into a bacterial plasmid, allowing multiple copies of the gene to be cloned.

14 D The heterozygous individuals exhibit insufficient cholesterol receptors, suggesting that neither allele is completely dominant nor recessive. There is no evidence of the condition being sex-linked.

15 C For the specified dose, C inhibited about 50% of viral entry and was relatively non-toxic to human cells.

16 B In DNA, adenine pairs with thymine and guanine pairs with cytosine. Adenine and guanine combining to make up 50% of nitrogenous bases is evidence of DNA being double-stranded (held together by the paired bases).

17 B SNPs that do not affect phenotype are likely to occur in the non-coding sections of DNA that are not expressed.

18 D The most likely suspect would have SNPs in common with both I and II. *D* is related to I (his grandmother is the sister of I's grandfather) and is also related to Il through *C* and *B*. *A* is unrelated to either I or II. *B* is the grandmother of II but unrelated to I. *C* is the son of *B* but unrelated to I.

19 A In the organism $2n = 6$, the key process in meiosis is the reduction division resulting from cells with n number of chromosomes. All four diagrams show chromosomes that have replicated or are made of two chromatids joined at the centromere. Only diagram A shows a phase that results in cells that each contain n chromosomes.

20 A Adoptive siblings that are genetically unrelated having similar traits (trait *A*) indicates a strong influence of the family environment. Identical twins having low similarity (trait *C*) suggests a low influence from genes on the trait.

Section II

Question 21 (Total 3 marks)

Three strategies that could prevent the spread of cholera include:

1 Treat all water used in food preparation or for consumption in order to kill bacteria. Chlorination or boiling should kill most bacteria.

2 Public health programs, such as handwashing, will improve hygiene in food preparation and sanitation. This may include promoting the cooking of vegetables rather than eating them raw.

3 Ensure safe disposal of human waste away from any water sources. (*3 marks*)

Question 22 (Total 3 marks)

Benefits of using pharmaceuticals such as antibiotics to treat infectious disease are quicker recovery of patients and decreased mortality of some infectious diseases.

A limitation of using pharmaceuticals is the increased likelihood of developing drug-resistant pathogens, making antibiotics less effective in the long term. (*3 marks*)

Question 23 (Total 3 marks)

(a) Point mutation (one nitrogenous base (T) substituted for another (A)) (*1 mark*)

(b) Chromosomal mutation which results from chromosomal rearrangement such as deletion, duplication or inversion of part of a chromosome (*2 marks*)

Question 24 (Total 7 marks)

(a)

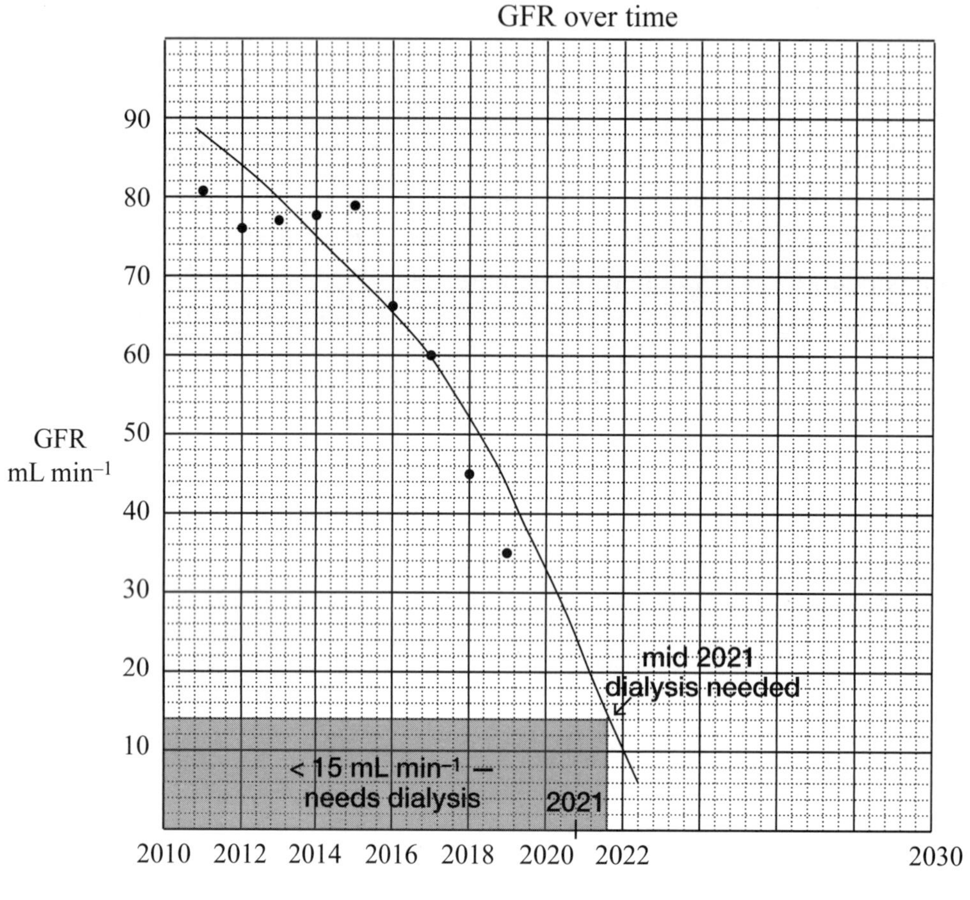

(*2 marks*)

(b) By extrapolation from the graph, dialysis will be needed in mid-2021. (*2 marks*)

(c) Dialysis compensates for the loss of functions of the kidney by the passive removal of nitrogenous wastes from the blood. Diffusion and osmosis allow the wastes to enter the dialysate from the blood. Electrolyte and water balance is achieved by controlling the composition of the dialysate. (*3 marks*)

Question 25 (Total 7 marks)

(a) You cannot draw a valid conclusion from the data because the sample of species used for comparison was too small and biased. There was failure to take into account the number of reproductive cycles in the lifetime of each species. The standard deviations were very large so that the difference in the values of the means was not significant. (*3 marks*)

(b) The hypothesis involved the testing of the number of eggs/young produced in animals using internal versus external fertilisation. The analysis of eggs laid in one reproductive cycle cannot be regarded as an accurate indication of the number of young produced over a lifetime as there can be great variation in the frequency of cycles, fertilisation success rates and the duration of fecundity. Taking into account the number of reproductive cycles in a lifetime would improve the experimental design.

Dramatically increasing the sample size and diversity of species sampled (such as including invertebrates as well as vertebrates) would improve the reliability of the data. (*2 marks*)

(c) External fertilisation is less demanding of the energy needed for parenting and allows more widespread dispersal of embryos. (*2 marks*)

Question 26 (Total 6 marks)

(a) Both parents were orange but half of their male offspring were yellow, indicating both parents carried a recessive allele for the yellow colour. (*2 marks*)

(b) If autosomal inheritance occurs, the yellow female parent (II) must be homozygous and the orange male parent (I) could be either homozygous or heterozygous. If (I) is homozygous, all offspring would be orange. If (I) is heterozygous, half the offspring would be yellow, irrespective of which gender the offspring were.

Sex-linked inheritance is detected because there is a link between gender and phenotype in offspring. If sex-linked, one gender would have two alleles for colour and the other would have only one. Assuming a pattern of sex-linked inheritance similar to human XX/XY sex-linked inheritance, all female offspring would be orange as they would inherit an X chromosome from the father carrying the dominant orange allele. All male offspring would be yellow.

X^o — X chromosome carrying the recessive yellow allele
X^O — X chromosome carrying the dominant orange allele

Punnett Square for sex-linked inheritance of fish colour (assuming male is XY, female is XX)

	X^O	Y
X^o	X^OX^o	X^oY
X^o	X^OX^o	X^oY

(*4 marks*)

Question 27 (6 marks)

(a) Controlling variables such as age, sex, education and socioeconomic status adds to experimental validity. Also contributing to validity was the calculation of each young adult's average exposure to arsenic at one of three levels and this was compared with survival rates for an 11-year period. The assessment of survival rates could be considered to be of reasonable duration and difference in exposure to contribute to a valid indication of mortality. (*2 marks*)

(b) The data appears to support the hypothesis because there is increased mortality of both males and females when there is increased average levels of exposure to arsenic in drinking water. Also the large sample size suggests that the study was reliable, although this may be improved if the study was carried out over a larger area. There is also no indication of the numbers of males and females studied over each level of arsenic exposure. Just the shape of the graphs might suggest a lot more participants fell into the lower arsenic bracket, bringing into question the graphs for mid and high levels of arsenic.

The data raises significant questions because of variations in the data between males and females. At the lowest levels of arsenic measured, females showed less mortality than the males over the duration of the data collection period. For the mid-level of average exposure the females had a higher mortality rate than the males, while at the highest levels of average exposure the patterns of mortality for males and females were similar. There was little variation in mortality rates of females experiencing mid and high levels of average exposure to arsenic. These factors suggest that further investigation is required before the hypothesis is unequivocally supported. Further investigation should focus on the types of diseases associated with the mortality of young adult males and females. Another area for further investigation is the impact of child-bearing on young adult females and their mortality due to arsenic exposure. Other gender-related factors that influenced mortality such as victims of violence across regions of the country need to be investigated.

While a large number of young adults were studied, there is no indication of the actual numbers that experienced low, mid and high levels of arsenic in drinking water. Even on the lowest point on the scale fewer than 1 in 1000 young adults died over the 11-year period. In the graph of female survival at the mid and high ranges particularly there were extensive periods (such as at least 2 years) in which no deaths were recorded. This suggests that there may have been few participants in these arsenic ranges, increasing the chances that variables other than average arsenic exposure through drinking water may have influenced survival. The fact that the lowest level of average arsenic exposure was up to nine times above WHO recommended levels raises concerns about whether this is a useful baseline for comparison of the survival rates at both higher levels. (*4 marks*)

Question 28 (6 marks)

(a) During the first phase of meiosis the chromosomes replicate to each become two chromatids joined by a centromere. The black paternal chromosome should have two black chromatids and the white maternal chromosome should have two white chromatids. Crossing over of the two adjacent chromatids will still result in recombinant chromatids but the diagram shows them with the wrong combinations of maternal and paternal chromosomes. (*3 marks*)

(b) Meiosis promotes genetic variation because of the random segregation of maternal and paternal chromatids and the formation of new combinations of alleles along the recombinant chromatids. Because meiosis results in haploid gametes it allows fertilisation to restore the diploid zygote and hence the genetic diversity resulting from the combination of alleles from two parents. (*3 marks*)

Question 29 (5 marks)

Mutation, gene flow and genetic drift (in a small population) can affect the gene pool of populations. Mutation increases genetic diversity because it results in new alleles. Mutations may also result in changes in gene regulation and expression or chromosome number and arrangement. Gene flow, the result of migration, also results in an increase in genetic diversity because not all subgroups within a population have identical gene pools. The gene pool (collection of all genes including the range of different alleles) of a population impacts evolution because survival and evolution is dependent on genetic diversity. Evolution results from natural selection. With greater genetic diversity a population is more able and likely to survive, reproduce and pass on characteristics to offspring which best equip them for their environment or changing environments. (*5 marks*)

Question 30 (7 marks)

Gene technologies impact on the management of infectious disease because they have been applied to facilitate the identification of various pathogens and infectious diseases such as the SARS virus. Gene technologies have been used for the diagnosis of the SARS-CoV-2 virus causing COVID-19. Tracing sources of infection has been made possible because of the detection of mutant strains in the virus. Testing sewage has also helped detect fragments of the virus to raise awareness of undiagnosed carriers of the disease in various locations and provide early warnings of potential outbreaks.

Gene technologies assist the production of safe vaccines to help prevent the spreading of an infectious disease.

Gene therapies have the potential to assist in the prevention and treatment of infectious diseases that resist conventional methods of treatment. This involves the introduction of genes that specifically block or inhibit gene expression or the functioning of gene products.

Some infectious diseases such as malaria are transmitted by vectors that carry the pathogens. CRISPR-Cas 9 and gene drive technology have the potential to disrupt the ability of particular species of mosquito to be able to carry the malaria pathogen.

Gene technologies have been successfully used in a range of non-infectious diseases. Genomic screening is used to detect a wide range of inherited diseases such as phenylketonuria. Gene therapy has the potential to impact management of inherited diseases such as cystic fibrosis and haemophilia by replacing faulty genes but it still requires research and clinical trials.

Genomic testing has been used to identify some risk factors for a range of non-infectious diseases such as breast cancer and some forms of cardiovascular disease. Single nucleotide polymorphism variations have been used as markers and associated with particular diseases and drug response effectiveness in various patients. Haplotype analysis has resulted in the need for only a small number of SNPs to be studied to map a single disease gene.

Pre-implantation genetic diagnosis has been used to screen embryos for in-vitro fertilisation to help select offspring without inherited diseases. Mitochondrial replacement therapy can be used to avoid transmission of genetic abnormalities carried on the female mitochondria.

Genetic engineering has been used to mass-produce human hormones such as insulin (for diabetes). Genetically modified organisms have been used to trial a range of medications for effectiveness and safety.

Gene technologies have been used to analyse the genome of tumour cells and so develop targeted cancer therapies. (*7 marks*)

Question 31 (9 marks)

(a) The data shows that blood glucose levels are influenced by diet, rest and the hormones insulin and glucagon. When the body is at rest the glucagon levels are relatively high but the glucose and insulin levels are low. At this point the hormone glucagon promotes the action of the liver to break down glycogen and release glucose so blood glucose levels are maintained. The large carbohydrate meal taken after one hour of rest results in an increase in glucose blood levels as digestion and absorption occur. Insulin levels rise after the meal, resulting in the liver taking up glucose and storing it as glycogen and the body cells taking up more glucose. By 45 minutes after the meal, the glucose levels start to fall. About one and a half hours after the meal, as the body metabolises glucose for energy the blood levels of glucose drop below a point that triggers the pancreas to release the hormone glucagon. This promotes the action of the liver to break down glycogen and release glucose so blood glucose levels stabilise.

Blood glucose levels are maintained within a range through a negative feedback loop in which the actions of the two hormones insulin and glucagon act in an opposing manner. Insulin stops the blood glucose rising excessively after a large carbohydrate meal and glucagon promotes the release of glucose into the blood after exercise or periods of fasting. (*6 marks*)

(b) Body temperature regulation is also controlled by negative feedback loops but the action of thermoregulation is through actions largely controlled directly by the nervous system (not the endocrine system as with homeostasis of glucose blood levels). Increased body temperature results in sweating and dilation of blood vessels close to the skin that allow cooling by evaporation, radiation and convection. A decrease in body temperature results in the nervous system constricting surface blood vessels and the shivering of muscles to result in heat production and increased body temperature. (*3 marks*)

Question 32 (18 marks)

(a) Adaptations of the rabies virus that facilitate transmission include its ability to survive and reproduce within an infected mammal vector. It must survive being secreted into the mammal's saliva before being transmitted to humans through a bite to successfully transmit from vector to host. The virus's ability to survive and replicate within human muscle tissue and then move into and within neurons and avoid the immune responses of the host maintains the successful infection and subsequent transmission of the virus. (*2 marks*)

(b) (i) The rabies virus cannot be classified as a cellular pathogen as it is unable to reproduce independently (other than in its host's cells) and does not have a cellular structure. It consists of a coat made of protein and lipid bilayer. Its internal RNA genome is contained within a protein coat and L and P proteins make up the RNA polymerase. It does not contain a cell membrane. (*3 marks*)

(ii) The role of viral RNA polymerase in the reproduction of the virus is as an enzyme that allows the transcription of the rabies virus's genome—single-stranded RNA—so that the polypeptide synthesis of its five proteins can occur within the host's cells. It allows the production of the appropriate mRNA so that translation into polypeptides can occur. Once the five proteins have been produced they can then be assembled into offspring rabies virus particles. L and P proteins make up viral RNA polymerase and diagram 2 shows them facilitating the production of complementary vRNA and then vRNA so that the essential components for viral assembly are present in the host cell. (*5 marks*)

(c) PEP works in two ways to help prevent the development of the rabies virus infection after exposure. Introduction of the HRIG antibodies will prompt their immediate reaction with the antigens on the rabies virus (the development of passive immunity). The antibodies join with the antigens, clumping them together so that they are more easily destroyed by the macrophages. Antibodies can also activate the complementary system that consists of a number of soluble proteins that circulate in blood and extracellular fluid. These proteins may directly destroy the viral envelope or destroy cells infected by the virus.

Artificially acquired immunity is achieved through the four injections of the rabies vaccine. This will trigger the immune system to produce antibodies. The booster injections serve to switch on the B lymphocytes that resulted from the first injection of vaccine to release antibodies more quickly and for a longer duration. Vaccines also promote the production of 'memory' T lymphocytes that remain to respond rapidly upon subsequent infection. The graphs demonstrate that the HRIG antibodies work instantly but decline over several weeks while the vaccine-induced antibodies take a week to increase in concentration but then remain high for an extended period.

The result of PEP is that there is no transmission of the rabies virus from the infected muscle through nerves to the central nervous system and hence the salivary glands remain free of infection. (*8 marks*)

CHAPTER 11

NSW Education Standards Authority

2021 HIGHER SCHOOL CERTIFICATE EXAMINATION

Biology

General Instructions

- Reading time – 5 minutes
- Working time – 3 hours
- Write using black pen
- Draw diagrams using pencil
- Calculators approved by NESA may be used

Total marks: 100

Section I – 20 marks

- Attempt Questions 1–20
- Allow about 35 minutes for this section

Section II – 80 marks

- Attempt Questions 21–33
- Allow about 2 hours and 25 minutes for this section

Section I

20 marks
Attempt Questions 1–20
Allow about 35 minutes for this section

Use the multiple-choice answer sheet for Questions 1–20.

---

1 A patient felt tired, weak and had a swollen neck. After following the doctor's advice to eat more foods containing iodised salt, her symptoms disappeared.

What was the most likely cause of the patient's symptoms?

A. Cancer

B. Genetic disorder

C. Nutritional deficiency

D. Environmental exposure

2 Which of the following photographs shows an example of sexual reproduction?

A.

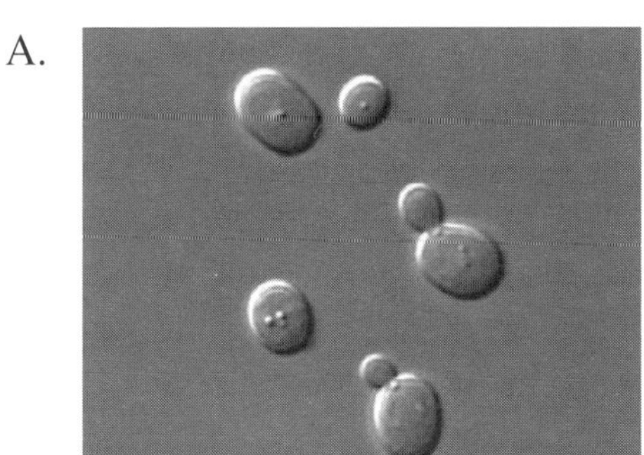

Yeast budding

B.

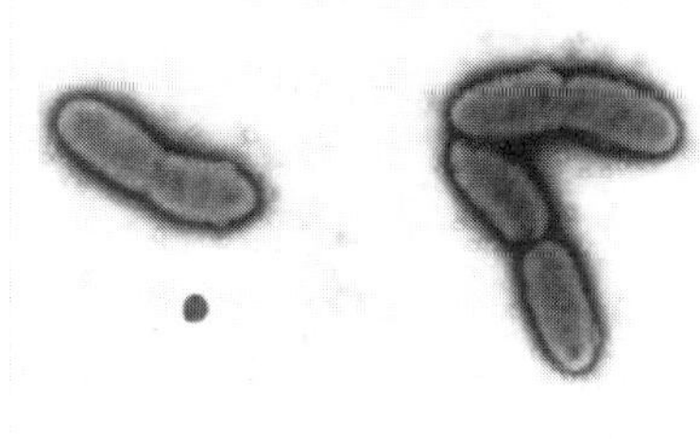

Bacteria – binary fission

C.

Strawberry runners

D.

Frog spawning

3 A scientist transferred male gametes from one plant to another to achieve a desired characteristic in the offspring.

Which genetic technology was the scientist using?

A. Gene cloning

B. Artificial pollination

C. Artificial insemination

D. Whole organism cloning

4 Which is the most effective strategy for treating non-infectious diseases?

A. Hygiene

B. Pharmaceuticals

C. Quarantine

D. Vaccination

5 Glucose levels are maintained by the hormones insulin and glucagon.

Which statement best describes the changes in hormone levels of a healthy human soon after a high glucose meal?

A. Insulin levels fall and glucagon levels rise.

B. Insulin levels fall and glucagon levels fall.

C. Insulin levels rise and glucagon levels fall.

D. Insulin levels rise and glucagon levels rise.

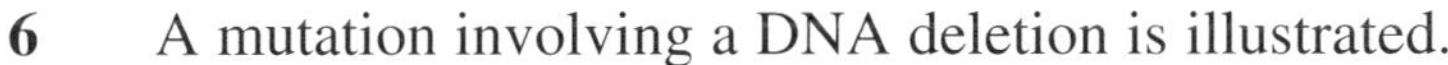

6 A mutation involving a DNA deletion is illustrated.

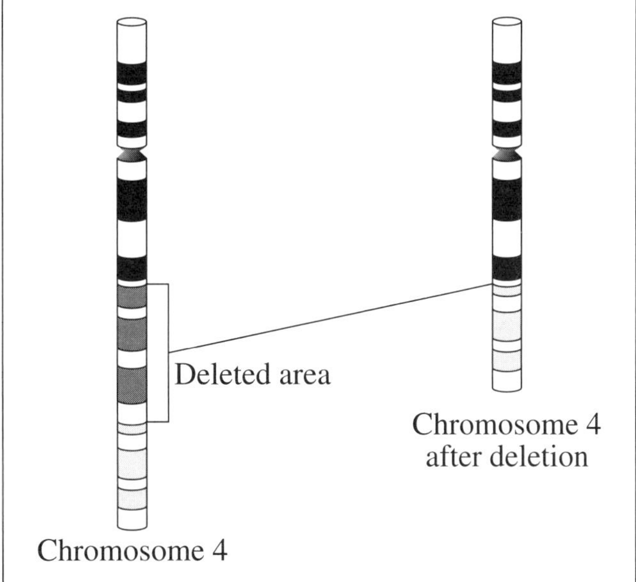

Which statement about the mutation is correct?

A. It will have an effect on many genes.

B. It will have an effect on only one codon.

C. It may be the result of an error during translation.

D. It may be the result of an error during transcription.

7 Which of the following shows correct base pairing in DNA replication?

A. A A T G T G C C A
U U A C A C G G U

B. A A T G T G C C A
G G C A C A T T G

C. A A T G T G C C A
T T A C A C G G T

D. A A T G T G C C A
T T U C U C G G T

8 Howard Florey conducted a breakthrough experiment in the development and use of antibiotics. He infected eight mice with *Streptococcus* bacteria. Four mice were given penicillin and survived while the four untreated mice died.

What conclusion could be drawn from the data obtained?

A. The experiment should be repeated with more mice.

B. The use of penicillin causes antibiotic resistance in mice.

C. Penicillin may be used on humans safely to treat bacterial infections.

D. Penicillin may have played a role in the survival of the four treated mice.

9 Streptomycin is an antibiotic that kills bacteria by interfering with the function of their ribosomes.

The primary effect of the antibiotic is that it prevents the bacteria from producing

A. tRNA.

B. mRNA.

C. amino acids.

D. polypeptides.

10 Cystic fibrosis is an autosomal recessive disorder caused by mutations in the *CFTR* gene. Many different recessive alleles cause cystic fibrosis.

The four most common alleles of the *CFTR* gene and their frequencies in the Australian population are shown in the table.

Allele	*Frequency of allele* (%)
A	98.33
a1	1.13
a2	0.08
a3	0.07

What will be the most common genotype of cystic fibrosis patients in Australia?

A. a1/a1

B. a1/a2

C. A/a1

D. A/A

11 Many transgenic crops have been genetically engineered to have traits such as herbicide resistance. In at least four different crops the transgene has been found in nearby wild plant relatives of the cultivated crops.

What is the most likely reason for this observation?

A. Crossing over in the wild plants

B. Gene flow from the crops to the wild plants

C. Genetic drift from the crops to the wild plants

D. Mutations in the wild plants that match the transgenes

12 The graph shows the levels of three hormones, oestrogen, progesterone and human chorionic gonadotrophin (HCG), measured in the blood of a woman during her pregnancy.

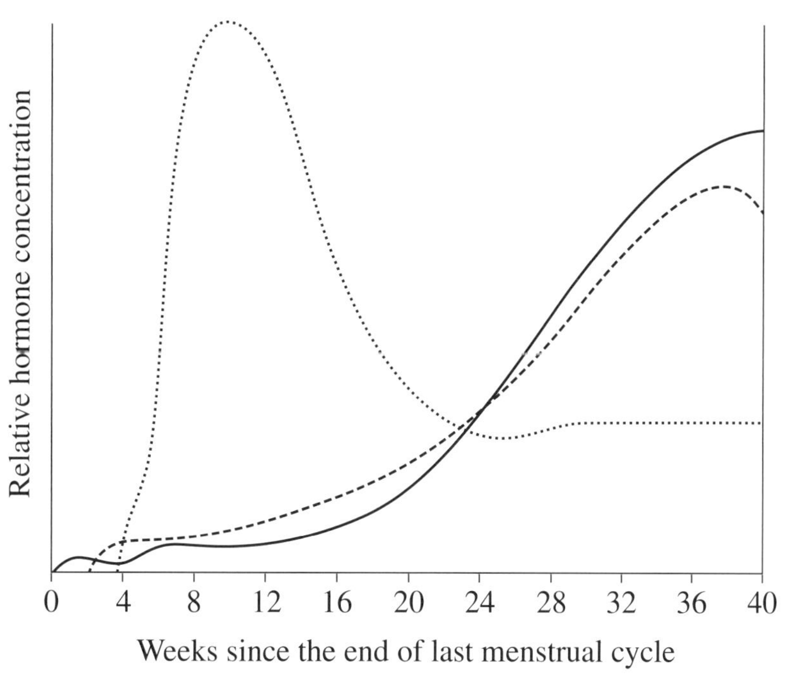

Which statement can be inferred from the graph?

A. Birth occurred about week 36.

B. Fertilisation occurred at day 0.

C. Implantation occurred about week 4.

D. The placenta was formed about week 24.

13 The photographs show an open and a closed stomate on a leaf surface. When open, stomates allow water vapour to pass out of the leaf.

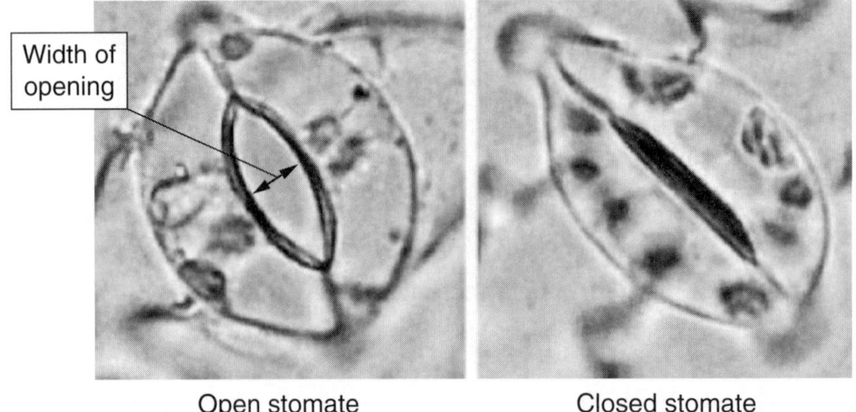

Regulating stomates is a mechanism by which plants maintain water balance.

Which of the following graphs best illustrates this homeostatic mechanism?

A.

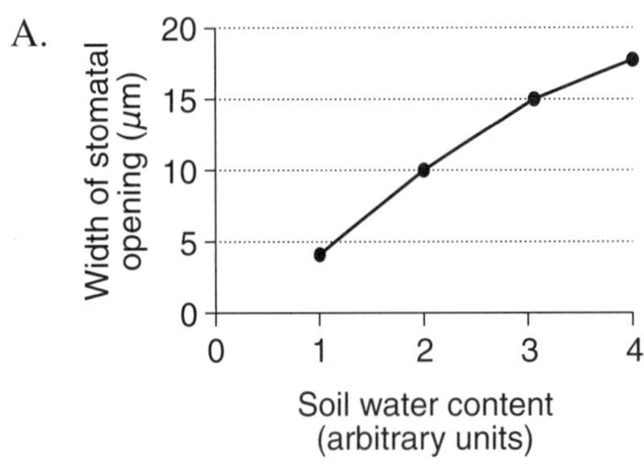

B.

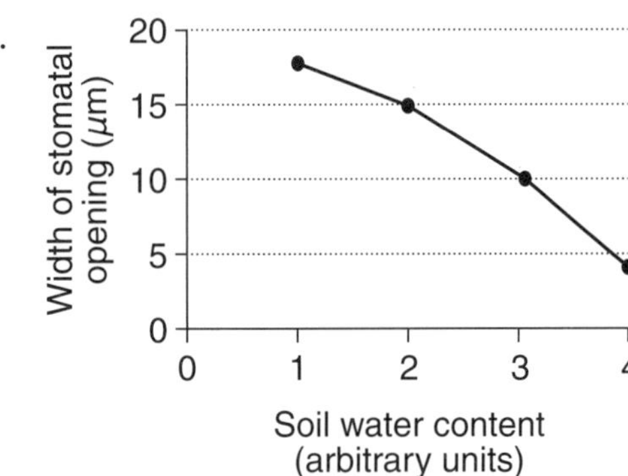

C.

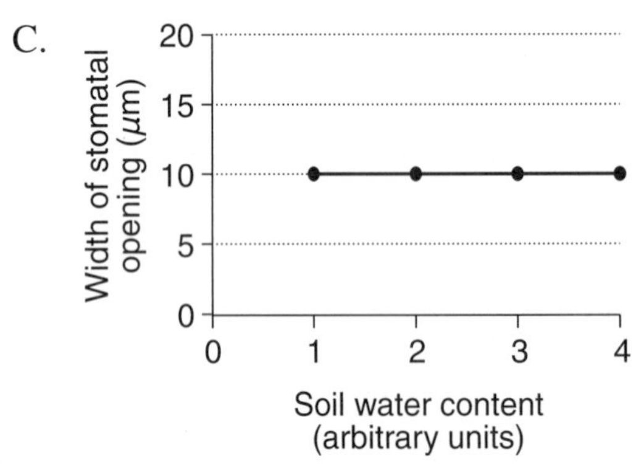

D.

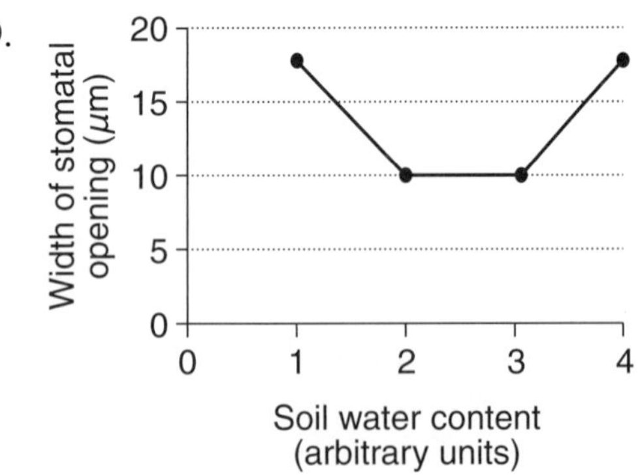

14 The human *PRNP* gene codes for the prion protein. Misfolding of this protein may be the result of ingesting tissue that contains misfolded prion protein or a mutation. Accumulation of misfolded prion protein causes serious diseases such as Creutzfeldt–Jakob disease (CJD).

Which of the following statements best classifies CJD?

A. It is both a genetic disease and an infectious disease.

B. It is a genetic disease only, since it is encoded by a gene.

C. It is not an infectious disease because the prion is non-cellular.

D. It is an infectious disease and the normal prion protein is the pathogen.

15 An example of the mutagenic effect of ultraviolet radiation (UV) on DNA is shown in the diagram.

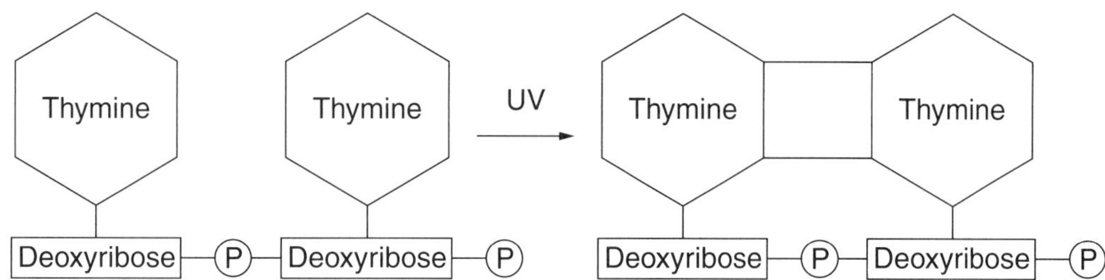

What is the mutagenic effect that is modelled?

A. Thymine is duplicated.

B. Bonds are formed between adjacent bases.

C. Nucleotides form bonds in the backbone of DNA.

D. Thymines on the two strands of DNA form bonds.

16 Scientists conducted an experiment to investigate the effectiveness of treating water from storage dams with UV radiation.

The experiment was conducted more than three times. The results are shown in the table.

UV dose (mJ/cm^2)	*Agar plates after being inoculated with 5 mL of water and incubated at 25°C for 24 hours*	
	Photo of agar plate	*Average number of bacterial colonies counted*
0		365
1		55
6		2

What conclusion may be drawn from the data obtained?

A. The control plates are contaminated.

B. High doses of UV eliminate all pathogens.

C. Exposure to UV inhibits reproduction of bacteria.

D. The presence of bacteria reduces the amount of UV.

17 The images show the sequence of changes in the chromosomes (stained black) during mitosis in plant cells.

Interphase
Prophase
Metaphase
Anaphase
Telophase

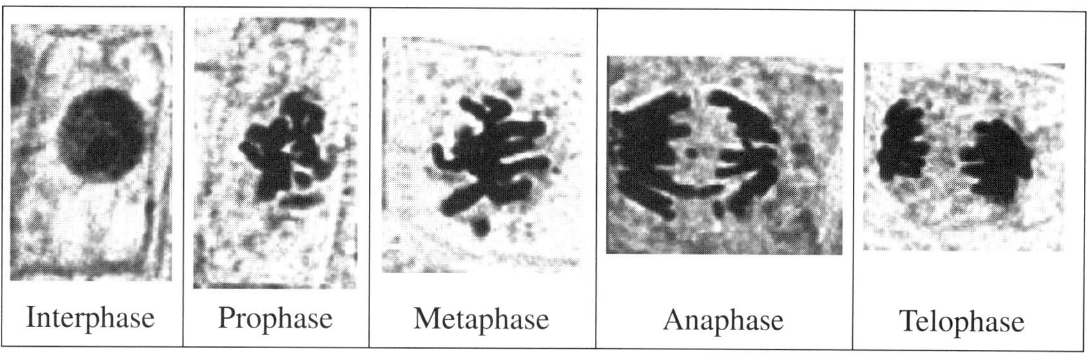

Which statement is true for mitosis?

A. Crossing over occurs during prophase.

B. Sister chromatids separate during anaphase.

C. Two daughter cells are produced during telophase.

D. Homologous pairs of chromosomes line up in metaphase.

18 Which human pedigree shows inheritance of a recessive, sex-linked characteristic?

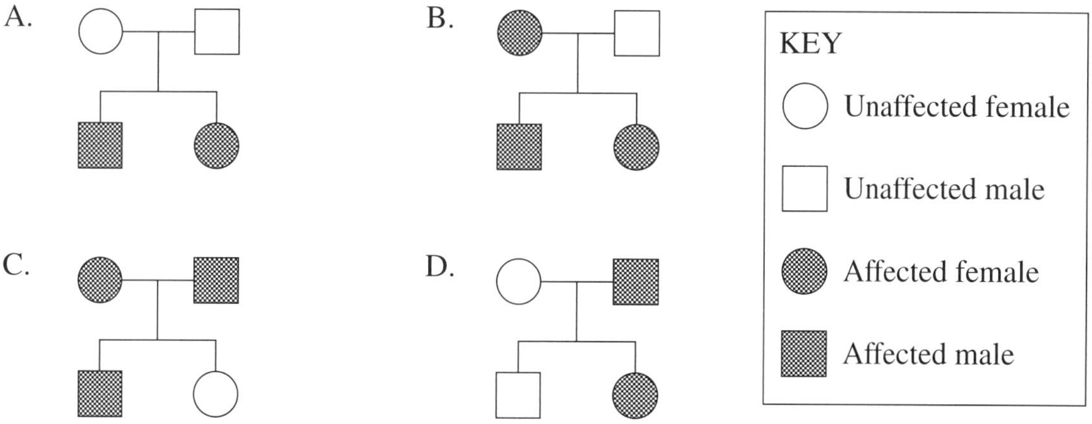

A.
B.
C.
D.
KEY
Unaffected female
Unaffected male
Affected female
Affected male

Use the following information to answer Questions 19–20.

Alcohol that is consumed is broken down by the following pathway.

Alcohol —(Alcohol dehydrogenase)→ Acetaldehyde —(Aldehyde dehydrogenase (ALDH))→ Acetate

ALDH is the product of the *ALDH* gene. The normal allele is *ALDH1* but many people have the *ALDH2* allele, which is associated with an increased risk of cancer.

Scientists have used mouse models to study the effects of *ALDH2*. They used wild-type mice with *ALDH1* alleles and genetically-engineered mice with one or two copies of the *ALDH2* allele.

The mice were given a dose of alcohol and the response was measured. The results are shown.

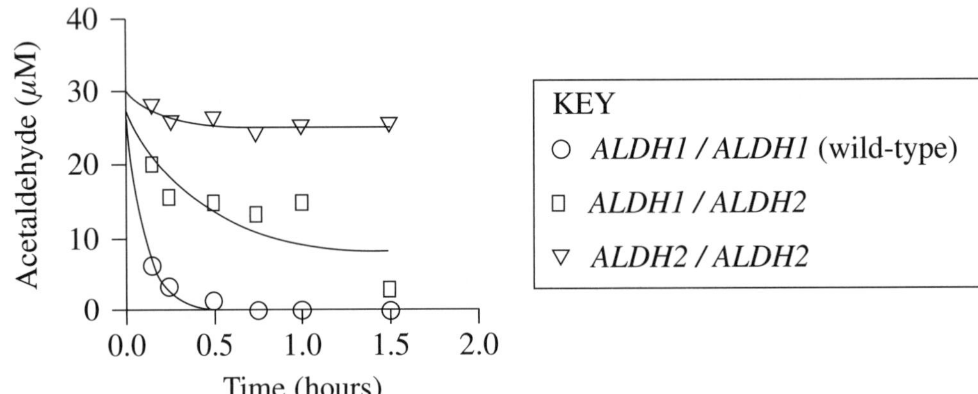

19 The *ALDH* alleles differ in their

A. genome location.

B. location in gametes.

C. effect on phenotype.

D. amino acid composition.

20 What do the data show about the effect of the *ALDH2* allele?

A. Enzyme activity is highest in homozygous mice.

B. Enzyme activity decreases if *ALDH2* is present.

C. Enzyme activity increases if *ALDH2* is present.

D. Enzyme activity is lowest in wild-type mice.

2021 HIGHER SCHOOL CERTIFICATE EXAMINATION

Centre Number

Student Number

Biology

Section II Answer Booklet

80 marks
Attempt Questions 21–33
Allow about 2 hours and 25 minutes for this section

Instructions

- Write your Centre Number and Student Number at the top of this page.
- Answer the questions in the spaces provided. These spaces provide guidance for the expected length of response.
- Show all relevant working in questions involving calculations.

Please turn over

Question 21 (7 marks)

(a) Label TWO features on the diagram below that would help to classify this pathogen as a bacterium. **2**

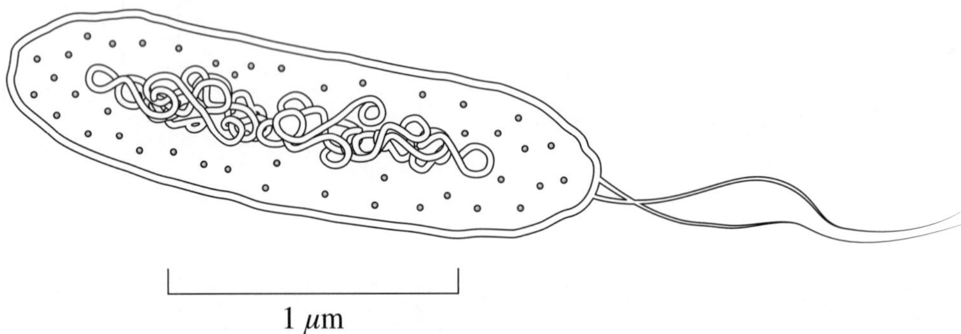

Question 21 continues on the following page

Question 21 (continued)

(b) A scientist followed Koch's postulates to confirm that this bacterium was causing diarrhoea in pigs on a local farm. **2**

Complete the boxes in the flowchart provided to show the steps taken by the scientist.

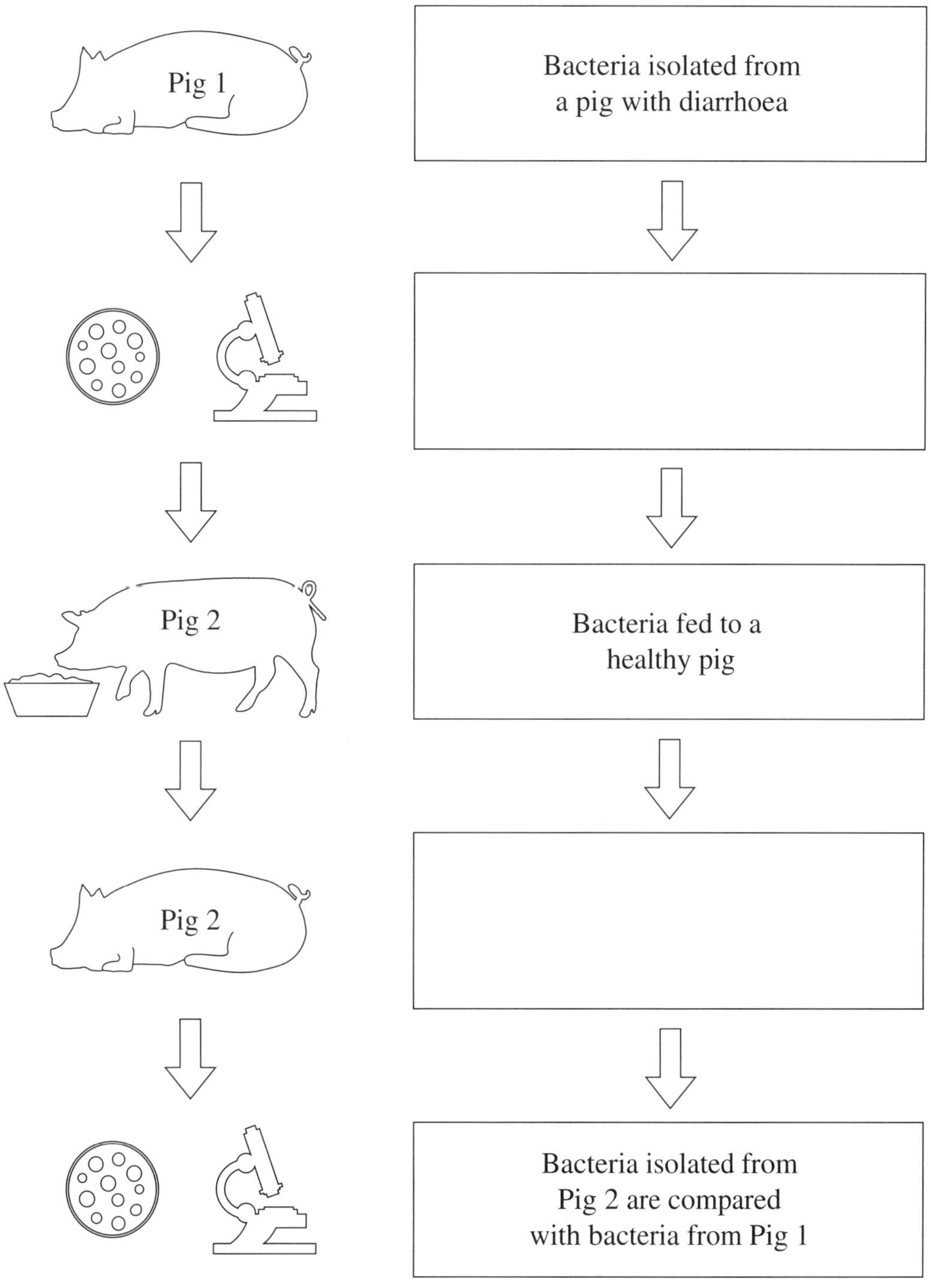

Question 21 continues on the following page

Question 21 (continued)

(c) Two pig farmers on neighbouring farms noticed that their pigs were suffering from diarrhoea and gradually losing weight. The farmers each adopted a different strategy to deal with this disease, as shown in the table. **3**

Farm	*Strategy*	*Result*
1	Treatment with antibiotics	All pigs recovered after two weeks
2	Elimination of rats and mice from pig sheds to improve hygiene	Decrease in number of sick animals over three months

Outline ONE benefit and ONE limitation of the strategies used on each farm.

....

....

....

....

....

....

....

....

Question 22 (3 marks)

In a population of rabbits, black fur colour is dominant over white fur. A black rabbit, whose mother has white fur, mates with a white rabbit. **3**

Predict the phenotypic ratio for the offspring of this cross. Show your working.

Question 23 (3 marks)

Ovulation in women is associated with a rapid increase in luteinising hormone (LH). Test strips can be used to detect high levels of LH in urine. Once a test strip is used, a control line should appear and the presence of a test line indicates high levels of LH in urine. The image below represents four different results that were obtained. **3**

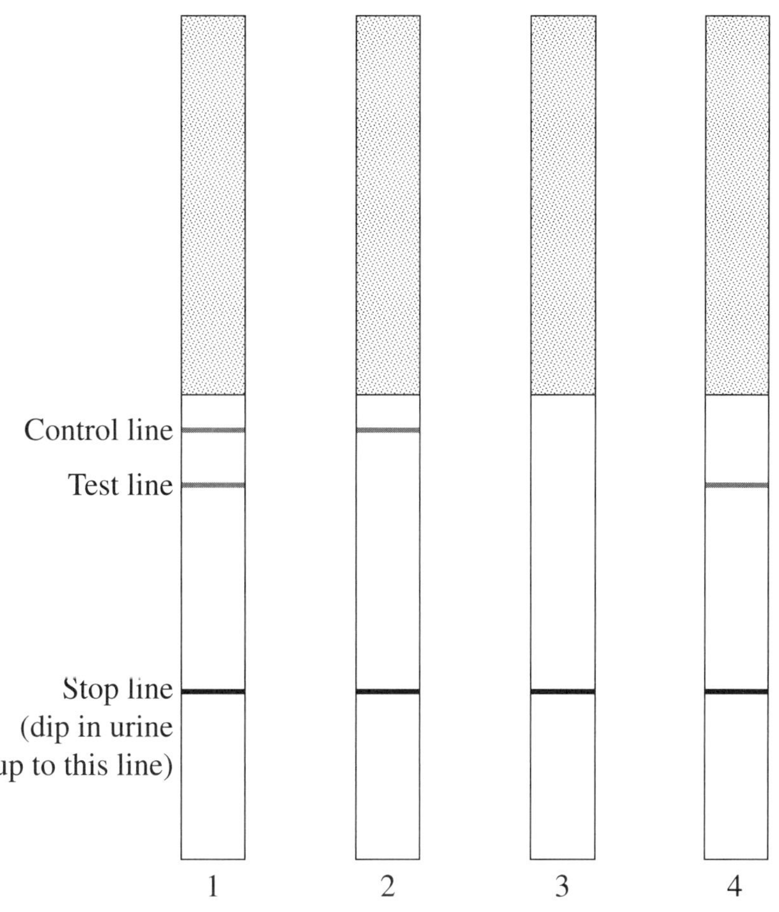

Which of the results indicates a valid test which shows that ovulation is NOT occurring? Justify your answer.

....

....

....

....

....

....

Question 24 (7 marks)

An incidence of an autosomal dominant trait is shown in the pedigree.

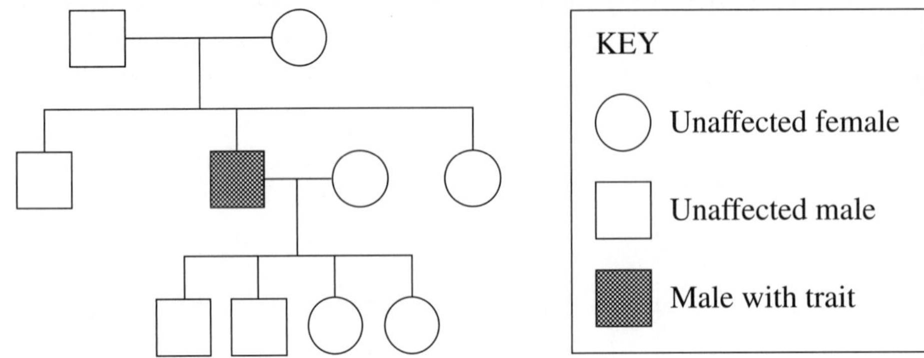

(a) Is this trait likely to be the result of a somatic or a germ-line mutation? Justify your answer. **3**

....

....

....

....

....

....

....

....

Question 24 continues on the following page

Question 24 (continued)

(b) The diagram shows the early stages of embryonic development from a fertilised egg. The developing ball of cells has split and monozygotic (identical) twins have formed. Mutations can occur at different times during embryonic development, for example Mutation *A* would result in both twins having the mutation in all their cells.

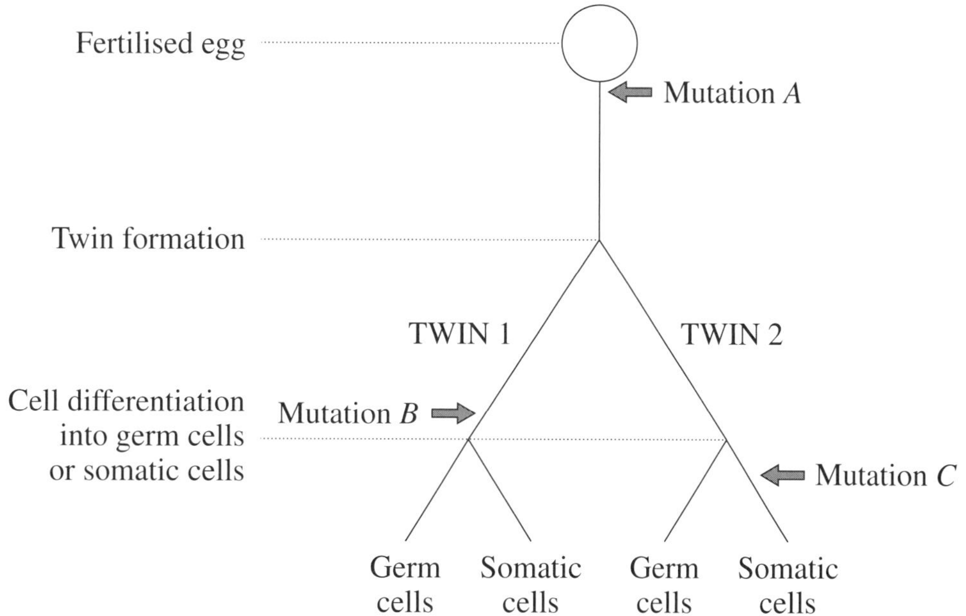

Explain the effects of Mutation *B* and Mutation *C* on each twin and on any offspring that they may have.

....

....

....

....

....

....

....

....

....

....

....

End of Question 24

Question 25 (8 marks)

A patient visited an audiologist for a hearing test. The audiologist tested both ears at specific frequencies. The volumes at which each frequency could be heard are shown.

Frequency (Hz)	*Minimum volume at which sound could be detected* (dB)	
	Right ear	*Left ear*
250	5	55
500	9	60
1000	9	75
2000	5	75
4000	9	80
8000	20	100

(a) Plot the data on the grid provided and include a key. **3**

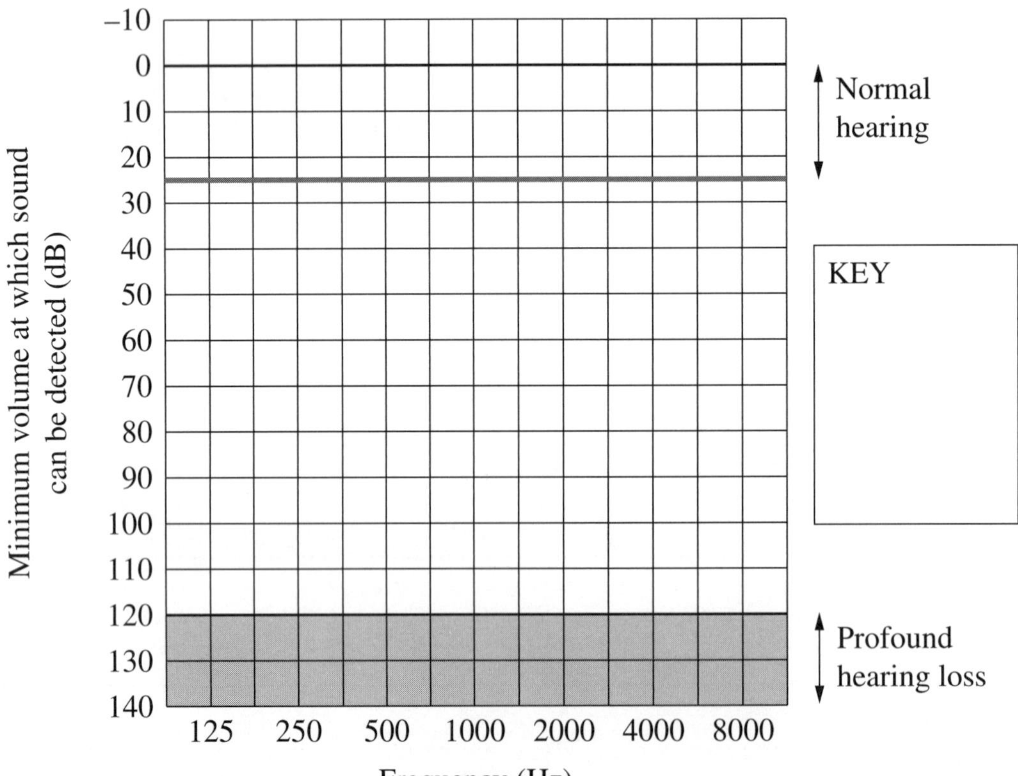

Question 25 continues on the following page

Question 25 (continued)

(b) What conclusions can be drawn about the patient's hearing?

....

....

(c) It is discovered that there is a complete and permanent blockage of the outer ear, but the cochlea is still fully functional.

Justify the use of a suitable technology to assist the patient's hearing.

....

....

....

....

....

....

....

....

End of Question 25

Question 26 (4 marks)

Zebra populations are suffering from a reduction in their gene pools due to habitat destruction and increasing isolation. This has led to an increase in the number of offspring born with coat patterns different to that of their parents. An example is shown. 4

Explain possible reasons for the increase in these offspring.

....

....

....

....

....

....

....

....

....

....

....

....

....

Question 27 (3 marks)

Sickle cell anaemia is a genetic disorder. In a family, the parents are both known to be heterozygous for the mutation that causes sickle cell anaemia. The couple has two unaffected children and is now expecting a third child. They have had an allele screening test to determine whether the child will have sickle cell anaemia. **3**

A part of the DNA profile is shown. It shows the alleles present.

Mother	*Father*	*Child 1*	*Child 2*	*Child 3*
▬▬▬	▬▬▬		▬▬▬	▬▬▬
▬▬▬	▬▬▬	▬▬▬	▬▬▬	

Use the DNA profile provided to justify whether Child 3 will have sickle cell anaemia.

Question 28 (8 marks)

(a) Describe the role of mRNA in human cells. **3**

....

....

....

....

....

....

....

....

Question 28 continues on the following page

Question 28 (continued)

(b) An mRNA vaccine has been developed in order to immunise people against a virus. The vaccine contains modified mRNA which codes for the spike protein on the surface of the virus. **5**

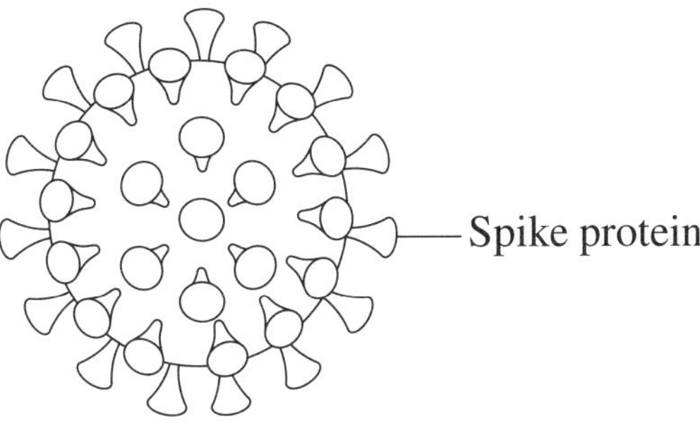

Explain how this vaccine can lead to active immunity to the virus.

....

....

....

....

....

....

....

....

....

....

....

....

....

....

....

....

....

....

End of Question 28

Question 29 (4 marks)

The koala is a mammal that maintains a stable body temperature of close to 36.6°C. **4**

A study was conducted. Koalas were observed in natural forests in south-eastern Australia. Their posture in the tree and the ambient temperature were recorded. Ambient temperatures were divided into two categories, hot and mild.

The graph shows the posture of koalas observed.

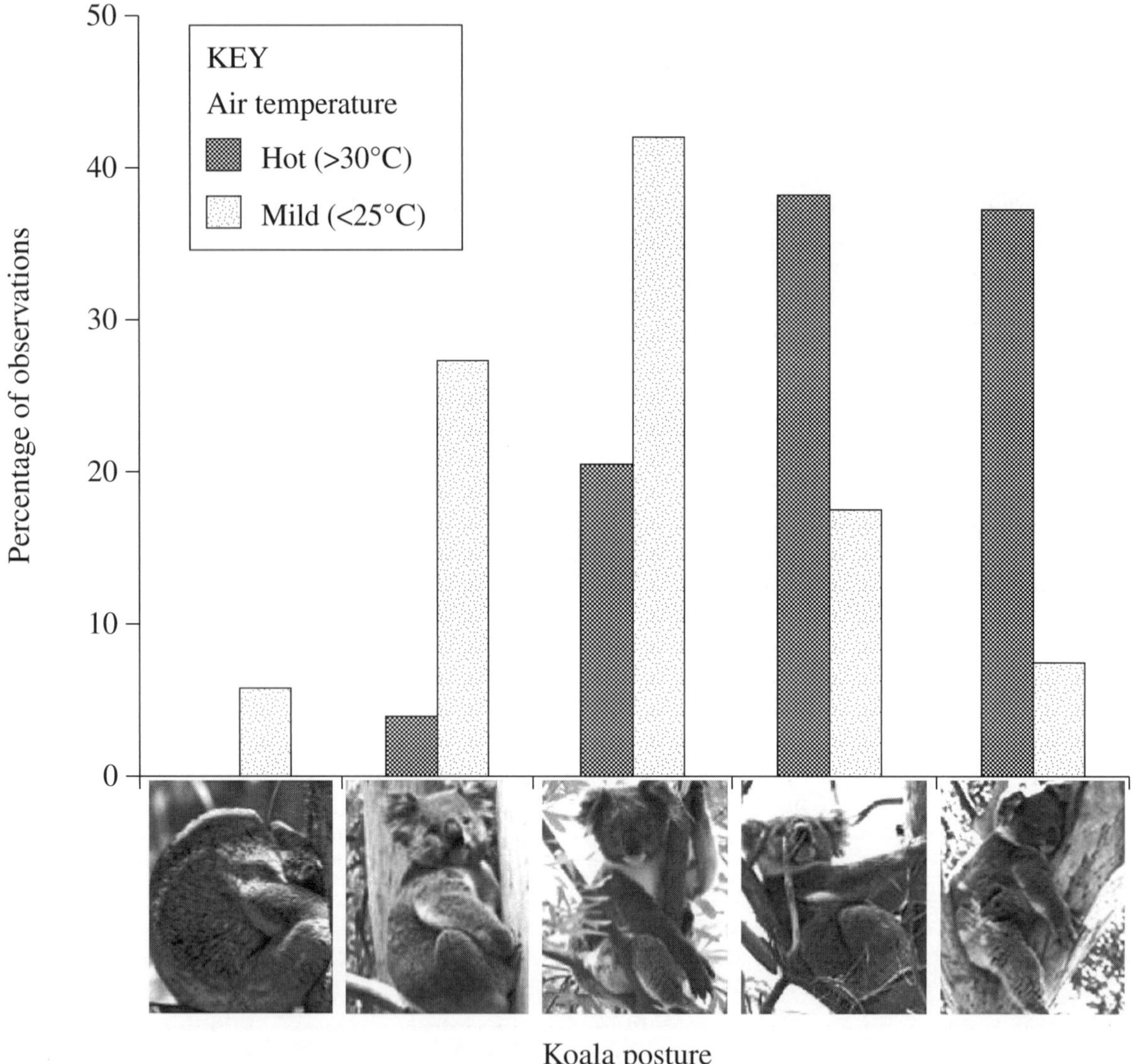

Additional data showed that the temperature of some tree trunks and large branches was up to 9°C cooler than air temperature during hot conditions.

Question 29 continues on the following page

Question 29 (continued)

Explain the adaptations used by the koalas in this study to maintain a stable body temperature. Make reference to the stimulus provided.

....

....

....

....

....

....

....

....

....

....

....

....

....

....

End of Question 29

Question 30 (7 marks)

A study compared the incidence of disease and survival of 8134 children who had received the measles vaccine with 8134 children from a neighbouring area who remained unvaccinated against measles. Children in each group were matched for age, sex, size of dwelling, number of siblings and maternal education. The graphs show the number of measles cases among the two groups over three years.

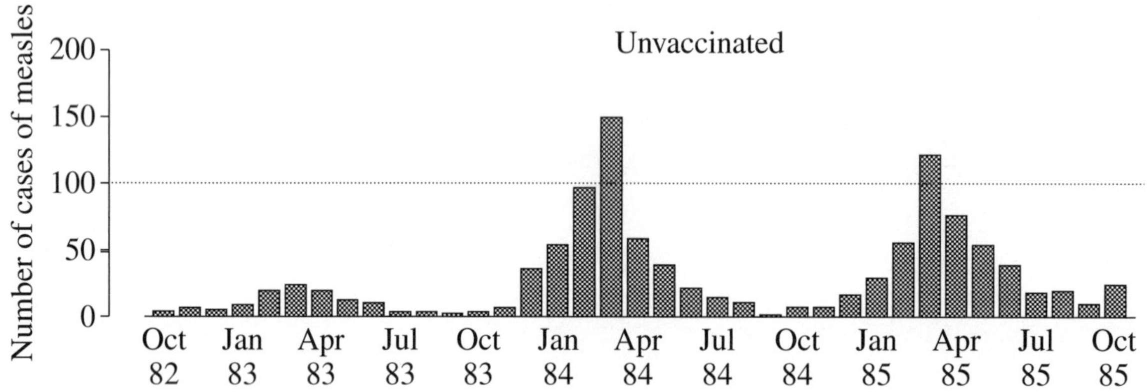

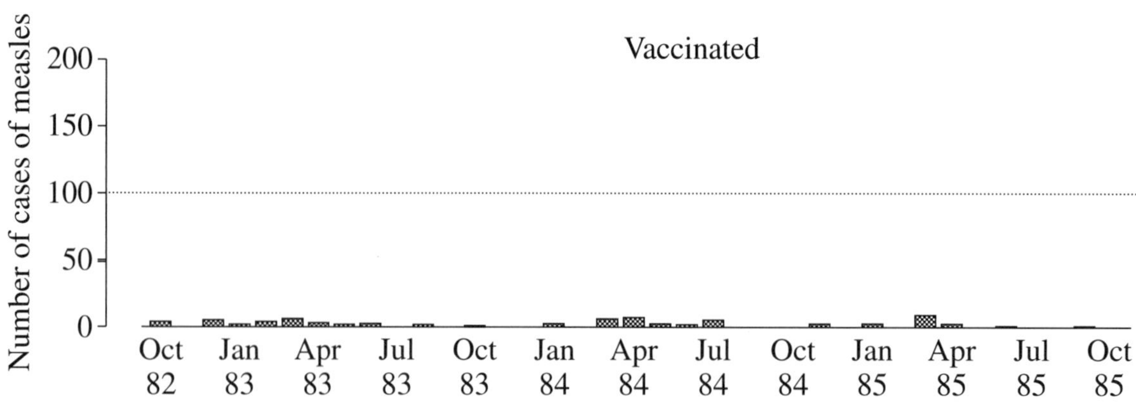

The table compares the cause of death and number of deaths of the two groups over the same three years.

	Number of deaths	
Cause of death	*Children vaccinated against measles*	*Children unvaccinated against measles*
Measles	2	40
Diarrhoea and dysentery	85	156
Oedema (swelling due to fluid in the tissues)	6	21
Fever	22	25
Total	**115**	**242**

Question 30 continues on the following page

Question 30 (continued)

'A vaccine only protects the community against a specific disease.'

Analyse the data with reference to this statement.

End of Question 30

Question 31 (6 marks)

Millions of people around the world take drugs known as statins, which have been shown to reduce the incidence of heart attacks and strokes in vulnerable patients. However, up to 20% of people stop taking statins due to side-effects such as muscle aches, fatigue, feeling sick and joint pain. **6**

A recent study at a public hospital focused on 60 patients who had all stopped taking statins in the past due to severe side-effects. Patients took statin tablets for four months, placebo tablets for four months and no tablets for four months.

Every day for the year the patients scored, from zero to 100, how bad their symptoms were. The results are shown.

Four month treatment	*Average score symptoms / 100*
Statin tablets	16.3
Placebo tablets	15.4
No tablets	8.0

Evaluate this study and its results.

Question 32 (5 marks)

The flow chart shows negative feedback by the hormones testosterone and inhibin in a human male. 5

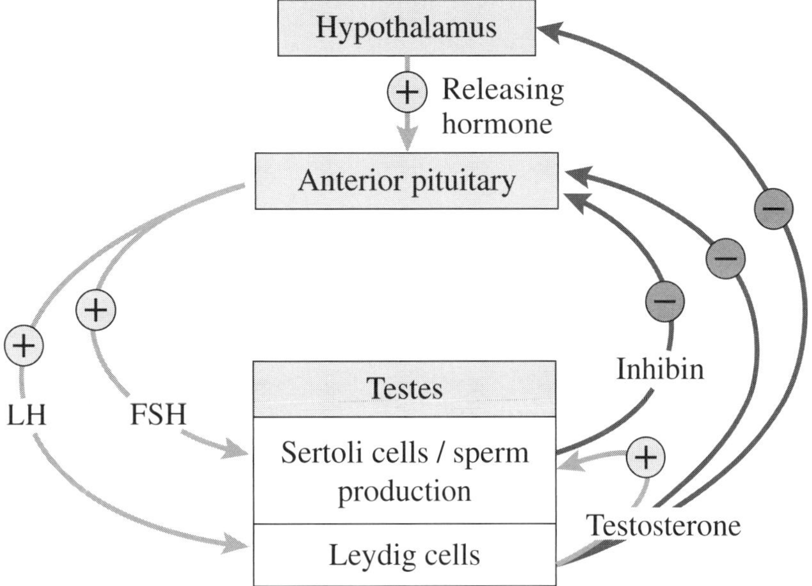

Some athletes take anabolic steroids to increase their muscle mass and strength. These steroids may be testosterone or a synthetic modification of testosterone.

Explain the changes that would occur in the testes of a male athlete continuously taking anabolic steroids. Support your answer with reference to the flow chart.

Question 33 (15 marks)

Genetically engineered Atlantic salmon have been produced and approved for aquaculture in the US. These salmon have a transgene that includes a protein-coding sequence from a Chinook salmon's growth hormone gene and the promoter region of an Ocean Pout's antifreeze protein gene. The following diagram provides an overview of the production of the transgenic salmon.

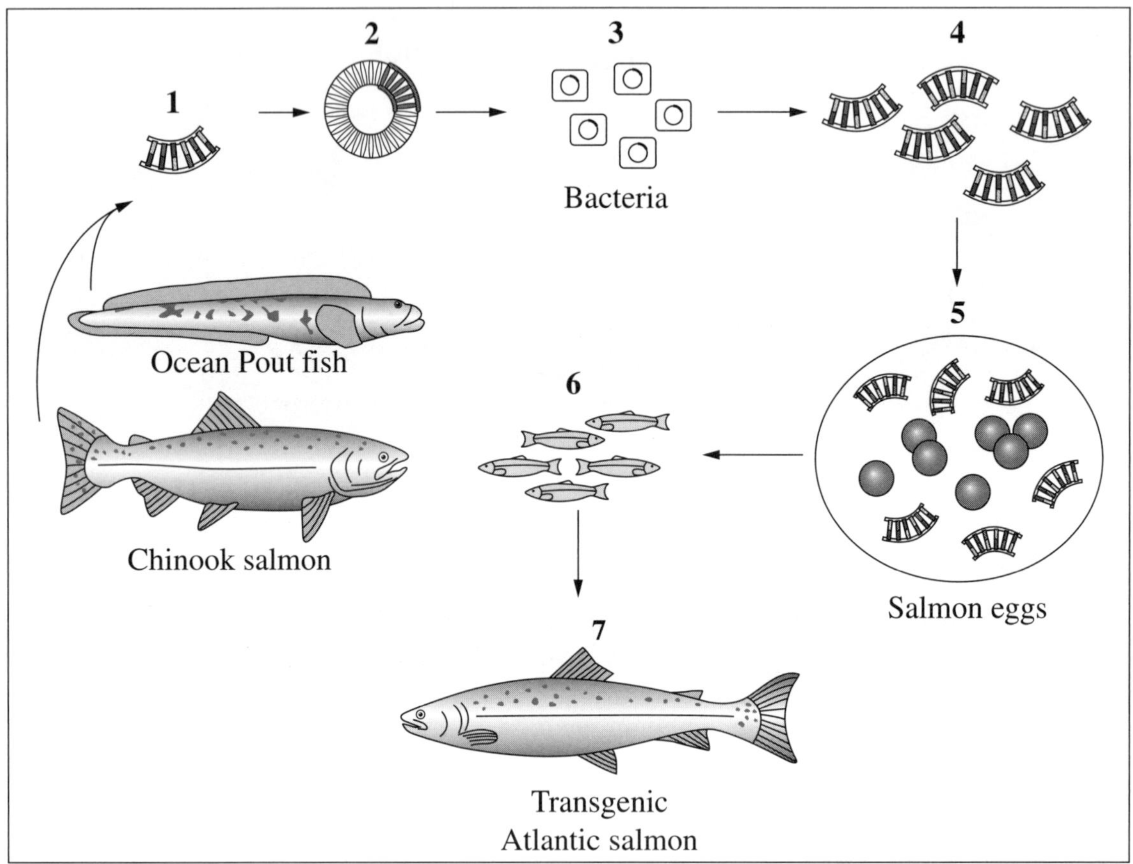

(a) Explain the processes shown in steps 1–4. **3**

....

....

....

....

....

....

Question 33 continues on the following page

Question 33 (continued)

(b) The graph summarises the growth of standard salmon and transgenic salmon. **3**

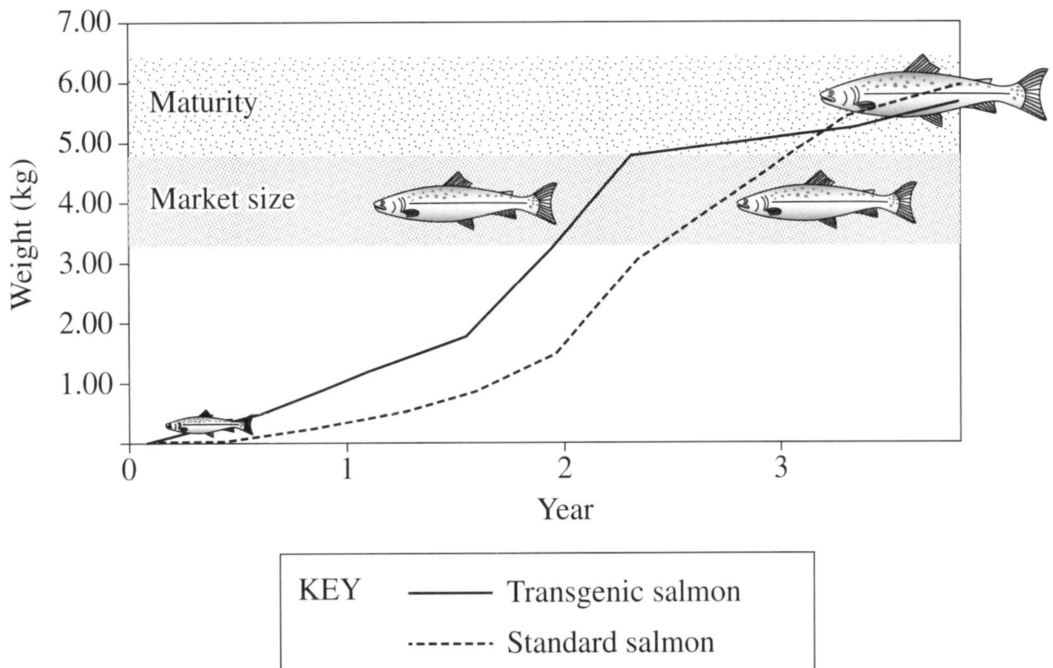

Explain ONE potential benefit of using transgenic salmon in aquaculture. Support your answer with data from the graph.

....

....

....

....

....

....

....

Question 33 continues on the following page

Question 33 (continued)

(c) Transgenic fish can reproduce and pass on the dominant transgene (T). **9**

Reproduction for aquaculture is strictly controlled using a variety of techniques in order to protect and preserve biodiversity.

Some of these techniques are outlined below.

1. Homozygous (TT) female (XX) breeding stock are kept in quarantine.
2. The female fish undergo hormone treatment that results in sex reversal and the development of male sex organs and sperm.
3. The sperm produced is collected and used to fertilise eggs obtained from wild-type, non-transgenic salmon.
4. The eggs are treated with pressure shock to prevent the completion of meiosis II. As a result, offspring are triploid (three copies of each chromosome).

   All offspring are transgenic female fish and have XXX (XXX fish cannot develop sex organs).
5. Offspring are transported to inland aquaculture tanks to be grown to market size.

Analyse how these techniques protect and preserve biodiversity.

....

....

....

....

....

....

....

....

....

....

....

End of paper

2021 HSC Examination Paper

Sample Answers

Section I (Total 20 marks)

1 C Iodine deficiency can be remedied with the increased intake of iodised salt. Iodine is essential for the functioning of the thyroid gland in the throat. It produces thyroxin that assists metabolism.

2 D Frog spawning is an example of sexual reproduction as it involves the production and union of male and female gametes.

3 B Transfer of male gametes (pollen) in flowering plants increases the chances of the offspring having desirable characteristics carried in the male genotype.

4 B Hygiene, quarantine and vaccination are effective strategies for treating infectious diseases. Pharmaceuticals may be of benefit for some inherited diseases, cancers and nutritional diseases.

5 C Insulin and glucagon have opposite effects on blood glucose levels. Insulin levels would rise to increase glucose storage and lower blood glucose levels. Glucagon levels fall so less glucose would be released from storage.

6 A Mutations caused by chromosome deletions are likely to impact many genes.

7 C DNA replication involves the pairing of A with T and G with C. U is only in RNA, not DNA.

8 D While it is important to replicate experiments with living organisms before drawing conclusions, the valid conclusion from the data is that the penicillin may have played a role in the survival of the four treated mice.

9 D The function of the ribosome is to facilitate the combination of mRNA with tRNA to link amino acids into a polypeptide chain.

10 A Cystic fibrosis patients are homozygous recessive so their most common genotype would involve the most common recessive allele a1.

11 B Gene flow results in the movement of genes between populations resulting in increased genetic diversity. Genetic drift results from random chance events impacting mainly small populations.

12 C HCG is mainly produced by the placenta so its levels increase after implantation and the development of the placenta.

13 A As water availability in the soil increases, the stomata can open more widely to allow greater transpiration.

14 A If the cause is a mutated gene then it is a genetic disease; if CJD results from ingesting infected material it is infectious.

15 B The diagram represents the formation of two bonds between adjacent thymine bases.

16 C Increased levels of exposure to UV radiation resulted in dramatically reduced numbers of bacterial colonies so it can be concluded that the radiation inhibited bacterial reproduction.

17 B A and D options relate to events in meiosis not mitosis. The image labelled anaphase shows the separation of sister chromatids. C shows telophase but must be followed by cytokinesis to result in the formation of two daughter cells.

18 D This pedigree can be explained by the unaffected mother being a carrier (only one X chromosome has the recessive allele). The daughter inherited two affected X chromosomes, one from each parent. The son would have inherited the unaffected X chromosome from his mother and so does not have the condition.

19 C The phenotype that is affected by the alleles is the rate of reduction of acetaldehyde.

20 B Homozygous *ALDH2* shows very little reduction in acetaldehyde and hence very little or no enzyme activity. The heterozygous condition shows a faster decline in acetaldehyde but not as much as in the homozygous wild type.

Section II

Question 21 (Total 7 marks)

(a) The size of approximately 2 micrometres is within the size range of bacteria, i.e. larger than a virus and smaller than Protozoan. Cell wall and plasma membrane surround the bacterial cell. DNA is coiled into a circular chromosome within the cell (not in a membrane-bound nucleus).See labelled diagram in photo. Two correct labels will score the marks. (Adding more information/labels is a problem if the information is contradictory.)

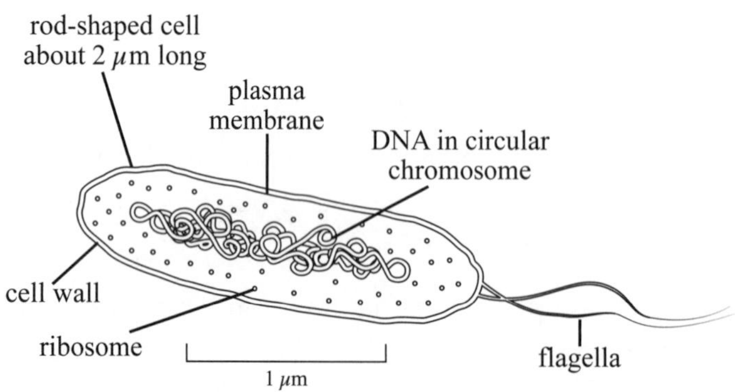

(2 marks)

(b)

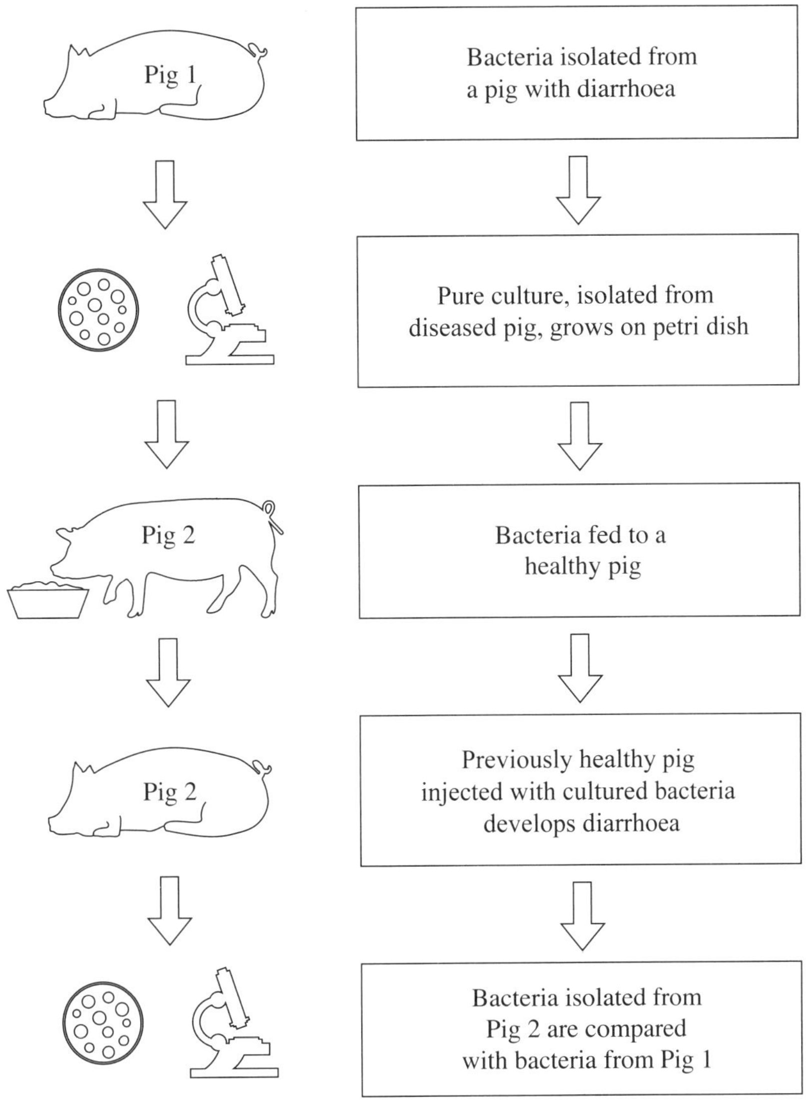

Pig 1
Bacteria isolated from a pig with diarrhoea
Pure culture, isolated from diseased pig, grows on petri dish
Pig 2
Bacteria fed to a healthy pig
Pig 2
Previously healthy pig injected with cultured bacteria develops diarrhoea
Bacteria isolated from Pig 2 are compared with bacteria from Pig 1

(2 marks)

(c) Treatment 1 has the benefit that the recovery is rapid and universal within the pig population at the farm. The limitation is that the bacteria causing the disease may develop antibiotic resistance so that subsequent attacks may result in less effective treatments and poor recovery rates.

Treatment 2 has the limitation that recovery took longer and was not universal through the pig population. The benefit is that long term the pigs at the farm will have greater resistance through the reduced chance of the bacteria becoming antibiotic resistant. Removal of rats and mice may also reduce the risk of pigs contracting other vector-transmitted infections or subsequent outbreaks. *(3 marks)*

Question 22 (Total 3 marks)

The phenotypic ratio of rabbit offspring would be one black (Bb) to one white (bb). The mated pair consists of a heterozygous black and a homozygous white.

Bb × bb
bb
bb
Bb
Bb
Phenotype ratio
1 black : 1 white
(Bb) (bb)

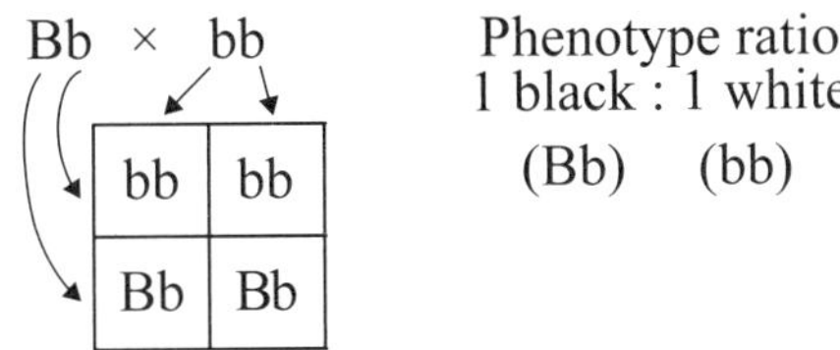

(3 marks)

Question 23 (Total 3 marks)

The test result that shows that ovulation is not occurring would not have high levels of luteinising hormone (LH).

Test result 1 can be discounted because the presence of a test line indicates high levels of LH.

Test results 3 and 4 appear to lack a control line which should appear if valid results are to be obtained. Test 4 could be further discounted because of the presence of a test line.

Test result 2 has a control line but no test line indicating that low levels of LH were present. This is a valid result demonstrating that ovulation is not occurring. *(3 marks)*

Question 24 (Total 7 marks)

(a) Germ-line mutations are generally associated with inherited conditions and passed on through generations. They have the potential to affect all body cells of the offspring. Somatic mutations occur in the body cells and are associated with diseases such as cancer and are generally not inherited.

As this condition is autosomal dominant and neither parents, siblings nor offspring have the condition it is unlikely to be a result of a germ-line mutation. *(3 marks)*

(b) Mutation *B* occurring before cell differentiation has the potential to affect somatic and germ-line cells in Twin 1 and so could become an inherited condition. It could affect all cells of Twin 1 and their offspring.

Mutation *C* occurs only in the somatic cells of Twin 2 so would not be passed on to subsequent generations but may affect Twin 2 throughout their life. *(4 marks)*

Question 25 (Total 8 marks)

(a) *(3 marks)*

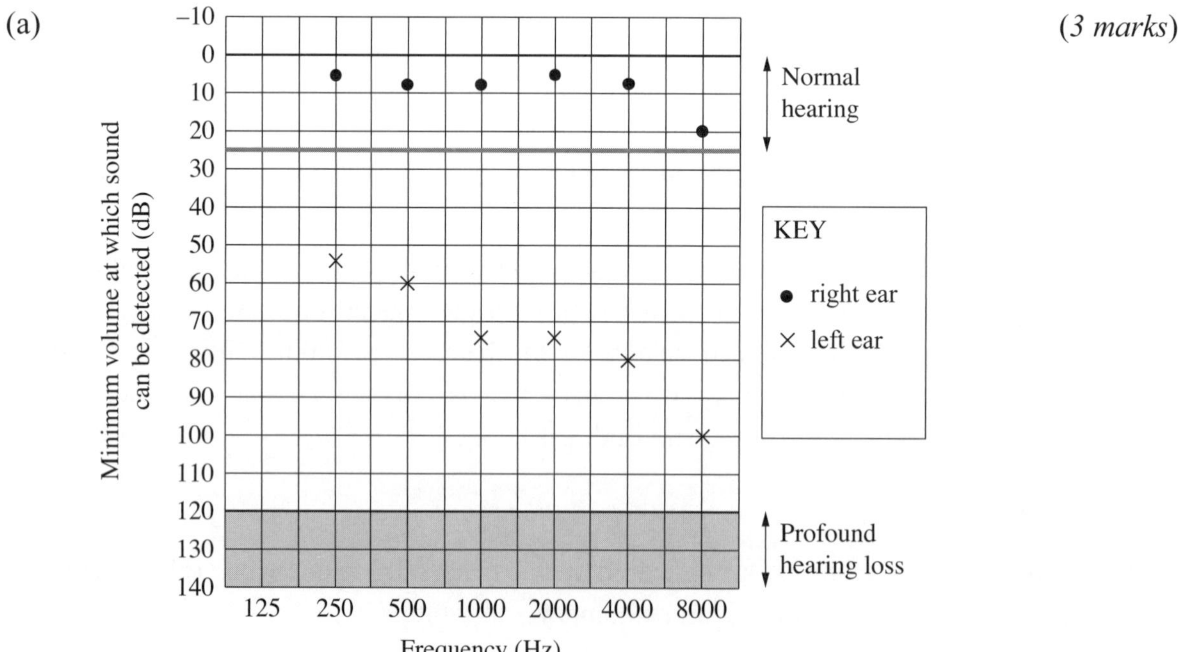

(b) The right ear had normal hearing across the frequency range tested. The left ear had impaired hearing which became progressively worse as sound frequency increased but was not at the level of profound deafness. *(3 marks)*

(c) A bone conduction hearing aid could be used to create vibrations that travel across the skull to the inner ear. The functioning cochlea would then transmit the vibrations so the brain could interpret the sound. The hearing aid would be worn behind the ear and a bone conductor vibrator fitted to a headband. This would address the cause of the problem: the blockage and conductive hearing loss. A normal hearing aid would amplify the sound vibrations which may not pass the blockage. A cochlea implant is unnecessary as the cochlea is functioning. *(3 marks)*

Question 26 (Total 4 marks)

Reduction in genetic diversity because of the pressures on the zebra population may result in more inbreeding. If the spotted coat pattern on the zebra was a result of a rare recessive allele, a small inbred population would be more likely to have the condition occur in the homozygous genotype and therefore the spotted phenotype would increase in frequency.

Genetic drift could be used to explain changes in allele frequency that occur in small populations as a result of random chance effects. Isolation of populations will prevent gene flow adding to genetic diversity. *(4 marks)*

Question 27 (Total 3 marks)

The allele present in Child 1 who does not have sickle cell anaemia must be the normal allele. It is also present in Child 2 and both parents, who are all without the condition. Child 2 has the same alleles as both parents and therefore must also be heterozygous. Child 3 does not have the normal allele so would have the sickle cell allele and the sickle cell anaemia. *(3 marks)*

Question 28 (Total 8 marks)

(a) mRNA plays a key role in the transcription of DNA in the cell nucleus to then carry the genetic code to the cytoplasm and ribosome so that the tRNA can facilitate the combination of amino acids in the correct sequence into a polypeptide chain. This process is important in human cells that are involved in secretion of enzymes, hormones or growing and repairing. *(3 marks)*

(b) The mRNA results in the production of a non-virulent spike protein that would be detected by the adaptive immune system consisting of antibodies, B cells and T cells that specifically target the pathogen. This artificially acquired immunity is primed to respond to exposure to the virulent form of the pathogen because it produces memory cells that bring about a faster, more concentrated response. The non-virulent spike protein acts as an antigen that binds to B cells. Helper T cells activate the B cells that differentiate into plasma cells that make pathogen-specific antibodies. Some B cells become memory cells.

Artificially acquired immunity also operates through the activation of the innate immune system. T cells and macrophages produce cytokines. Some cytokines inhibit viral replication. Other cytokines act on the acquired immune system as hormone-like messenger substances or activate other immune cells. *(5 marks)*

Question 29 (Total 4 marks)

The koala images in hot and mild conditions demonstrate a range of behavioural adaptations to temperature change. The posture of the koala rolled into a ball (on the left) and apparently asleep was only recorded in mild conditions. This posture would reduce the surface area of the koala over which heat would dissipate, thus helping to maintain a body temperature higher than the ambient temperature.

The koala in the photo second on the left shows the alert koala sitting upright with legs drawn in and making little contact with the trunk. This posture occurs more frequently in milder conditions. This koala would not lose body heat to the trunk or larger branches and folded legs would help retain body heat.

The middle photo of the upright and alert koala shows it has arms and legs wrapped around a smaller trunk. This posture is the most frequent for koalas experiencing mild conditions but is not uncommon in hotter conditions. The recumbent koala posture in the photo second from right is most common in hot conditions but not uncommon in mild conditions. Both these koalas may derive some cooling benefits from either the small trunk or large branch.

The posture of the sleeping koala on the right shows it is almost hugging the large trunk. As the trunk's temperature is up to 9 degrees cooler than the ambient temperature this would help cool the koala in hot conditions. Being inactive in hot conditions would also help the koala maintain body temperature at 36.6 degrees. (*4 marks*)

Question 30 (Total 7 marks)

The graphs demonstrate that the measles vaccine is very effective in reducing the incidence of measles especially at times when there are outbreaks of measles in the unvaccinated population. The data in the table also shows the vaccine is very effective in preventing deaths from measles. The vaccine protects the community from a specific disease (measles).

This table also shows a reduced death rate in vaccinated children from two other conditions but no significant difference in death rates from fever deaths between the two groups. The unvaccinated comparison group was selected from a neighbouring area. It is possible that the almost double rate of death from dysentery and diarrhoea in the unvaccinated population could have been a result of poor sanitation and water treatment in their neighbourhood or another uncontrolled variable between the studied populations.

The data also demonstrates that death from oedema was almost three times more frequent in the unvaccinated population. Drawing a causal link between vaccination and prevention of death from oedema is far from certain. Build-up of fluid in the tissues sufficient to cause death could possibly be the result of failure of the kidneys that are responsible for removal of nitrogenous wastes and also water and salt balance. It is possible that the matching of the vaccinated and unvaccinated groups was inadequate and there were dietary or genetic differences in the populations. If the two neighbourhoods were isolated there could be factors such as genetic drift.

Fever is a result of the inflammatory response to pathogens. The similar death rate from fever in both groups would suggest vaccination from measles gives no added support for the immune response to pathogens other than those causing measles.

The data is not conclusive in establishing whether vaccination against measles protects the community from other diseases. (*7 marks*)

Question 31 (Total 6 marks)

The study: This exploratory study is not a scientific investigation with a clearly stated aim and method. It has reliability concerns because of the small sample size and lack of replication. Many factors could have underpinned the participants' daily evaluation of their symptoms on a scale of zero to 100. Such a vast scale would suggest a high level of sensitivity that probably did not exist in such subjective decisions about symptoms such as fatigue and nausea. The fact that the participants took tablets or no tablets in sequence over 12 months could mean that factors such as weather, diet and activity levels may have affected symptoms and reporting. Ideally such a study would involve larger numbers of participants and controls.

The results: The average score of 8 while participants took no tablets suggests the participants had some ongoing health issues similar to the side effects of statins. The lack of a substantial difference between the reported side effect levels of the placebo and the statins suggests that concerns other than the side effects may have played a role in the participants deciding to no longer continue taking statins. The reported results are also questionable as the average scores out of 100 are relatively low for patients supposedly experiencing severe side effects. The range of results recorded, or even the changes in individual scores rather than just averages, may give more insight. *(6 marks)*

Question 32 (Total 5 marks)

The diagram shows that the Leydig cells in the testes naturally produce testosterone that has a positive impact on Sertoli cells and sperm production. Negative feedback results in the suppression of the hypothalamus and anterior pituitary because of the production of inhibin by the cells and the increased testosterone. The reduced activity of the anterior pituitary results in less activity of the Leydig cells, and reduced testosterone and sperm production.

Artificially increasing testosterone would have a similar impact on Sertoli cells and sperm production. The impact of taking anabolic steroids would be increased production of inhibin that would suppress the action of the anterior pituitary and hence reduce the production of FSH and LH. Taking anabolic steroids in the short term would act in a similar manner to increased levels of testosterone.

The long-term impact of using anabolic steroids would be to reduce the activity of the pituitary and anterior hypothalamus so that eventually the natural production of testosterone would decline. Within the testes activity of the Leydig cells, Sertoli cells and sperm production would decline.

Negative feedback ensures the body returns to homeostasis despite the use of anabolic steroids. *(5 marks)*

Question 33 (Total 15 marks)

(a) Steps 1 to 4 are essentially gene cloning. The two DNA segments, one promoter from Ocean Pout fish and one gene from Chinook Salmon, will be identified, and removed from the chromosomes of these fish. These segments of DNA will be combined to produce a functional unit, the promoter and the gene together. New technologies such as the CRISPR-Cas9 facilitate this process of identifying and gene editing. Then the new gene and promoter is inserted into a bacterial plasmid. Once the new gene unit is inserted into a bacterial plasmid it is transferred into a host bacterial cell where it replicates so that there are many copies of the transgene. *(3 marks)*

(b) The graph shows a steeper line for growth of transgenic salmon than the standard salmon. The transgenic salmon grow more rapidly than standard salmon, reaching marketable size about 4 months earlier than the standard salmon. This growth rate would have economic benefits for salmon farmers with an earlier, shorter period of harvesting. *(3 marks)*

(c) Biodiversity relates to genetic diversity within a species, between species in ecosystems and within ecosystems. Protecting and preserving species biodiversity in ecosystems involves taking measures to ensure transgenic species are not released into the wild. If this was to occur the more rapid growth of the transgenic species might result in them out-competing wild species and in the long term reducing species diversity. Another possibility is that they may interbreed with wild species and introduce transgenes into wild populations. This genetic pollution may initially increase genetic diversity within the wild species, but the long term impact on ecosystem biodiversity may not be predictable.

Keeping the transgenic female breeding stock in quarantine and their offspring contained within inland aquaculture should greatly reduce the risk of transgenic salmon entering the ocean and resulting in genetic pollution. Inland aquaculture has the added benefit that excess food, fish wastes and use of antibiotics or other artificial disease prevention methods can be contained more easily and not drain into oceans to impact more widely on other marine ecosystems.

Aquaculture can have a devastating impact on biodiversity if the sources of fish food are harvested in a manner that removes fish populations indiscriminately and/or destroys marine environments. Attempts to include plant-based food sources need to consider impacts on biodiversity in the areas where the plants are produced or harvested.

The use of hormone treatment to bring about sex reversal of female transgenic salmon allows sperm to be collected and then used to fertilise wild-type non-transgenic salmon eggs. Using hormones to reverse the sex of the female salmon will not change their sex chromosomes so that all sperm produced will carry an X chromosome. When this sperm is used to fertilise wild-type eggs only female offspring will result. This process has the added benefit that it ensures separation of transgenic and wild salmon so that wild salmon cannot naturally interbreed with or be exposed to transgenic salmon.

Pressure shock treatment of the eggs used to fertilise transgenic sperm ensures the transgenic offspring are triploid and therefore without sex organs. Being sterile will guarantee that there can be no transgenic salmon produced as a result of natural reproduction of transgenic fish. *(9 marks)*

CHAPTER 12

Excel

NSW Education Standards Authority

2022 HIGHER SCHOOL CERTIFICATE EXAMINATION

Biology

General Instructions

- Reading time – 5 minutes
- Working time – 3 hours
- Write using black pen
- Draw diagrams using pencil
- Calculators approved by NESA may be used

Total marks: 100

Section I – 20 marks

- Attempt Questions 1–20
- Allow about 35 minutes for this section

Section II – 80 marks

- Attempt Questions 21–32
- Allow about 2 hours and 25 minutes for this section

Section I

20 marks
Attempt Questions 1–20
Allow about 35 minutes for this section

Use the multiple-choice answer sheet for Questions 1–20.

---

1 A healthy person in a hot environment measures their body temperature to be 38.0°C.

Which of the following might occur in this person?

A. Shivering

B. Vasodilation

C. Goosebumps

D. Pale appearance

2 Some desert mammals can obtain their water from breaking down stored body fat.

What type of adaptation is this?

A. Behavioural

B. Environmental

C. Physiological

D. Structural

3 What type of protein is formed in response to a pathogen?

A. Antibody

B. Antigen

C. Antihistamine

D. Antiseptic

4 The diagram shows the response to an injury in a human.

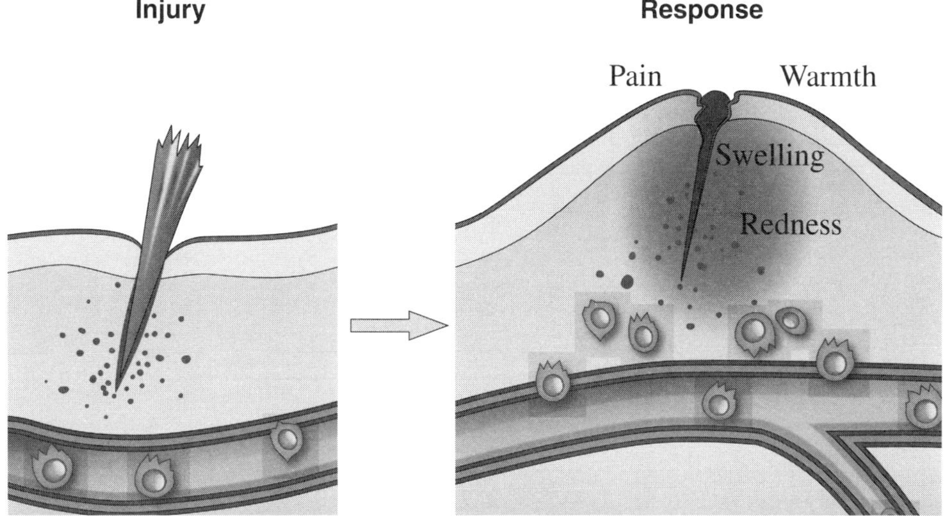

What is the response shown?

A. Infection

B. Inflammation

C. Phagocytosis

D. Vasoconstriction

5 Data for incidence of notifiable infectious diseases in Australia are shown in the table.

Disease	*Percent of all cases of infectious disease notified for 2019*
Influenza	53
Sexually transmitted infections	25
Gastrointestinal diseases	10
Vaccine preventable diseases excluding influenza	9
Other	3

Which type of graph should be chosen to best represent these data?

A. A column graph because percentages are continuous.

B. A column graph because diseases are categorical.

C. A line graph because percentages are continuous.

D. A line graph because diseases are categorical.

6 A diagram of the human female reproductive system is shown.

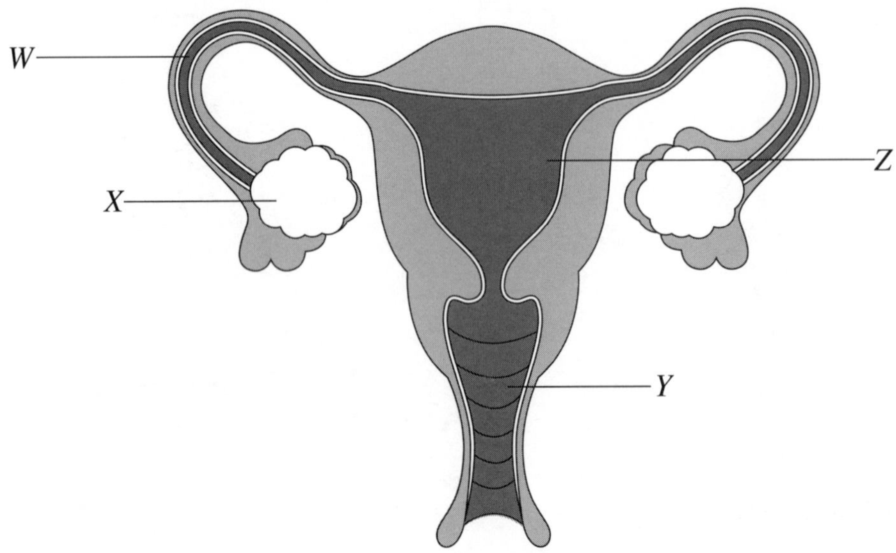

Where does implantation normally occur?

A. *W*

B. *X*

C. *Y*

D. *Z*

7 Animal cloning involves inserting the nucleus from a somatic cell of one animal into the

A. plasmid of a bacterium.

B. uterus of a surrogate animal.

C. fertilised egg from another animal.

D. enucleated egg from another animal.

8 In mice, muscle cells contain 40 chromosomes.

Which of the following is a correct statement about mouse cells?

A. The gametes contain 20 pairs of chromosomes.

B. The somatic cells contain 20 pairs of chromosomes.

C. The cells produced by meiosis contain 40 chromosomes.

D. The cells produced by mitosis contain 40 pairs of chromosomes.

9 A new vaccine against an infectious disease was developed. The effectiveness of the vaccine in preventing infection in humans was plotted over time in three different age groups.

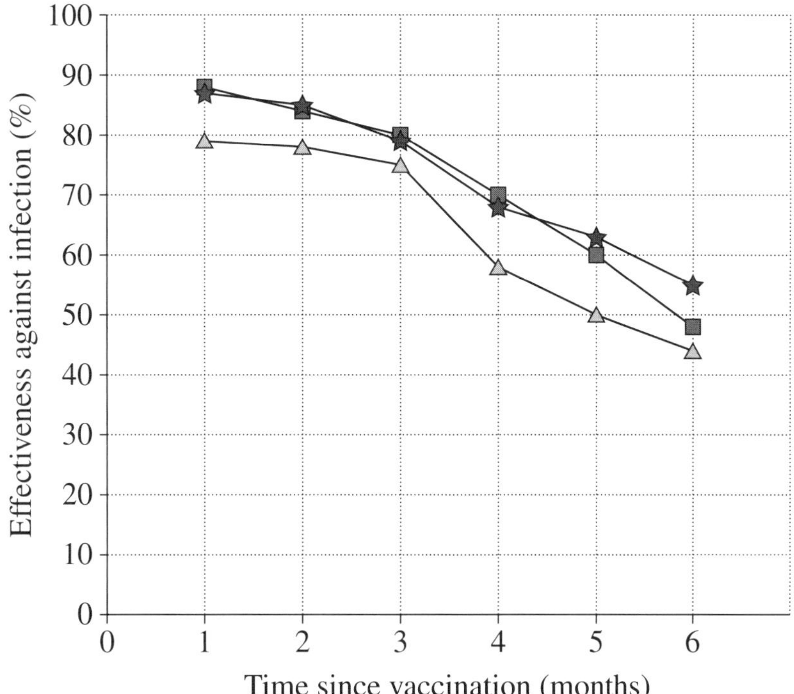

KEY

16–44 years — 45–64 years — ≥ 65 years

Which of the following is a valid conclusion that can be drawn from the data in the graph?

A. The vaccine is ineffective after six months.

B. The younger the person the more effective the vaccine.

C. The younger the person, the faster the immunity to infection declines.

D. The vaccine offers more protection to younger people than those over 65.

Refer to the following information to answer Questions 10–11.

Pasteur used swan neck flasks to conduct experiments on microbial contamination of broth. One of Pasteur's investigations is shown.

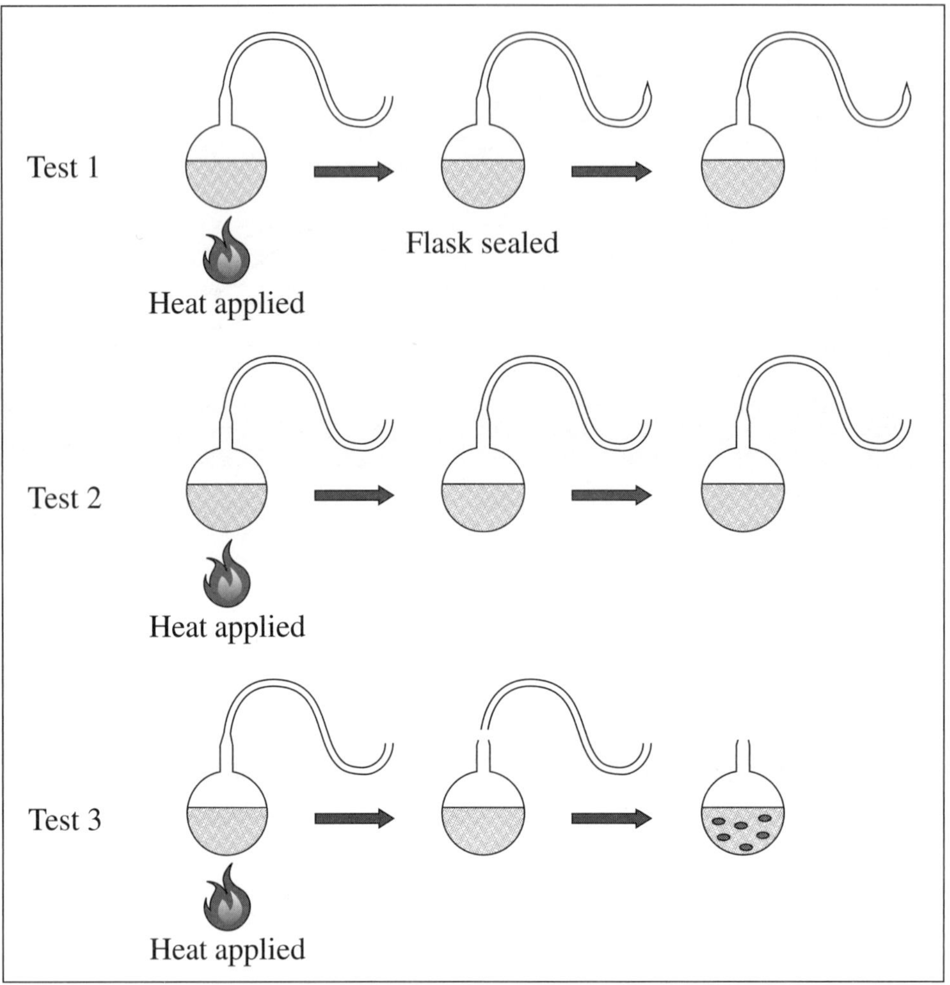

10 Which of the following was the independent variable in this investigation?

A. The air

B. The flask

C. The broth

D. The microbes

11 What is the best explanation for Pasteur's results?

A. Cells arise from existing cells

B. Heating prevents broth spoiling

C. Gases in the air cause broth to spoil

D. Cells arise by spontaneous generation

12 The following diagram represents a short section of DNA.

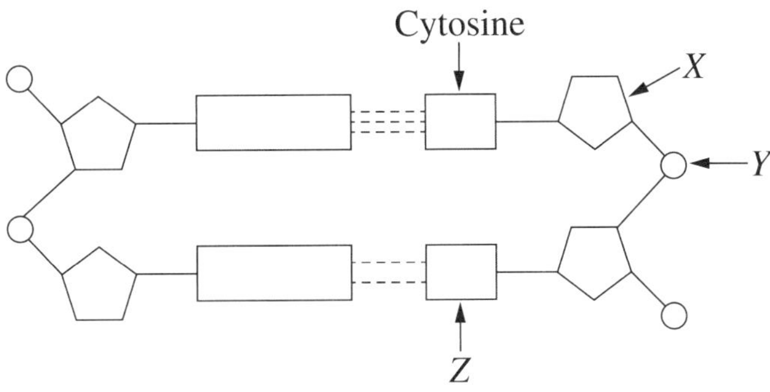

What are the components labelled *X*, *Y* and *Z*?

	X	*Y*	*Z*
A.	Deoxyribose sugar	Phosphate	Thymine
B.	Deoxyribose sugar	Phosphate	Guanine
C.	Phosphate	Deoxyribose sugar	Guanine
D.	Phosphate	Deoxyribose sugar	Thymine

13 The section of DNA shown in Question 12 came into contact with a mutagen, resulting in a change in the cytosine. The effect on the DNA after several replication cycles is shown.

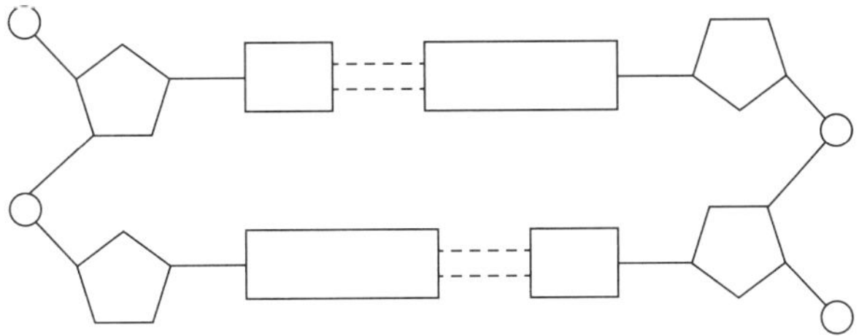

The mutagen has resulted in

A. a frame-shift mutation.

B. an increase in cytosine.

C. a change in one base pair.

D. a change in one nucleotide.

14 An inherited characteristic in a family is mapped in the pedigree shown.

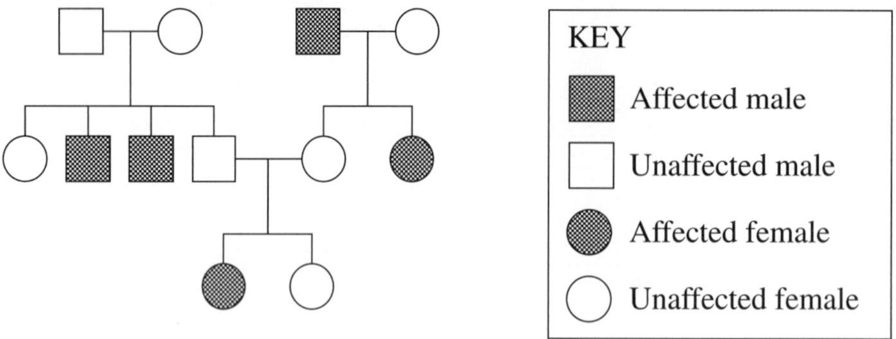

Inheritance of this characteristic is

A. autosomal recessive.

B. sex-linked recessive.

C. autosomal dominant.

D. sex-linked dominant.

15 In a plant species, red flower colour (R) is dominant over white flower colour (r).

Two plants of known genotype for flower colour were crossed. A punnet square was used to determine the proportion of genotypes expected in the offspring. Part of the punnet square is shown.

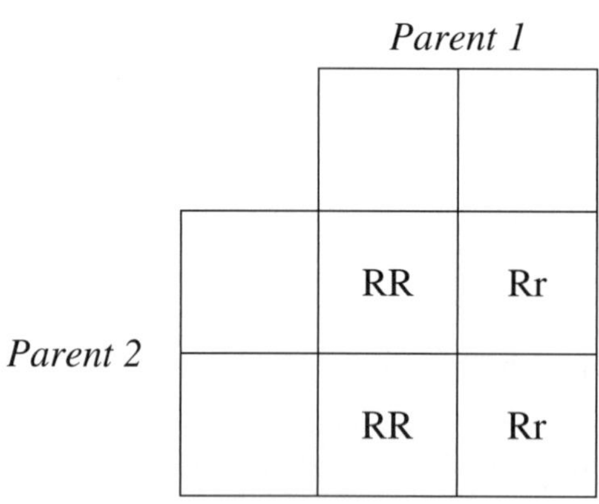

Which statement is true for the parents in this cross?

A. Both parents were homozygous.

B. Both parents were heterozygous.

C. Both parents had flowers of the same colour.

D. Parent 2 must have red flowers and Parent 1 must have white flowers.

16 What is the significance of mutations in non-coding DNA?

A. They do not alter nucleotides.

B. They do not affect phenotype.

C. They may affect gene expression.

D. They may alter amino acid codons.

17 A student constructed a model of the phases of meiosis in an organism with six chromosomes in its somatic cells. One of the early phases is shown.

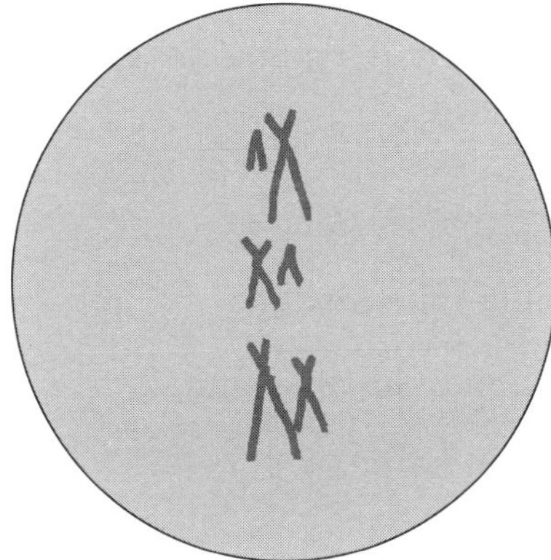

Which of the following identifies an error in the model?

A. Chromosomes are not lined up single file on the metaphase plate.

B. Chromosomes of matching size and structure are not paired.

C. There are no homologous chromosomes.

D. There are too many chromosomes.

18 The occurrence of a genetic disease in a family resulting from the presence of a dominant allele is shown.

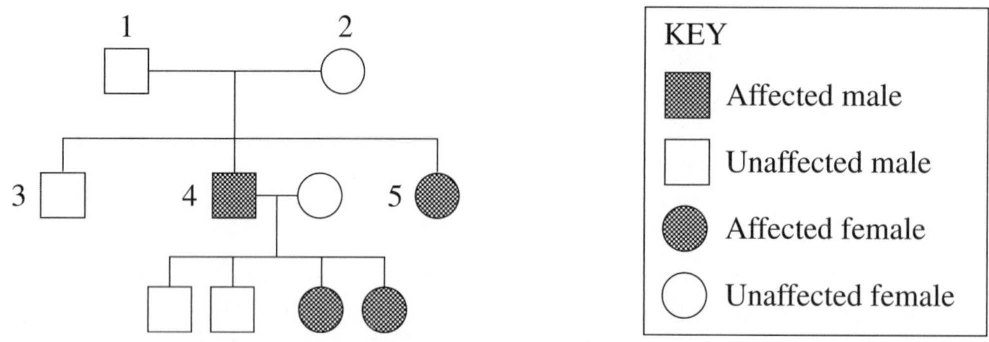

If the disease has arisen as a result of a mutation, which of the following is the most likely cause of the disease in this family?

A. A germ-line mutation in individual 2

B. A germ-line mutation in individual 4

C. A somatic mutation in individuals 1 and 2

D. A somatic mutation in individuals 4 and 5

19 A person with short-sighted vision (myopia) was incorrectly given a pair of spectacles with convex lenses.

Which row of the table demonstrates what happens?

	Without spectacles	*With spectacles*
	Distant object — Eye	Distant object — Lens — Eye
A.	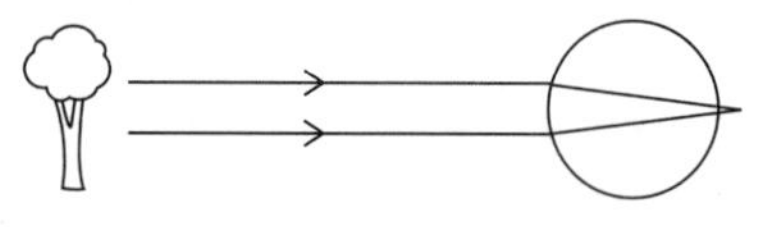	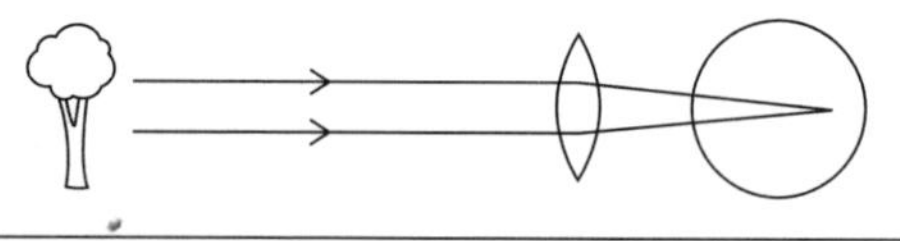
B.	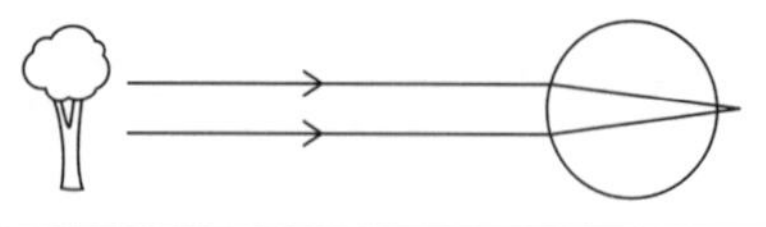	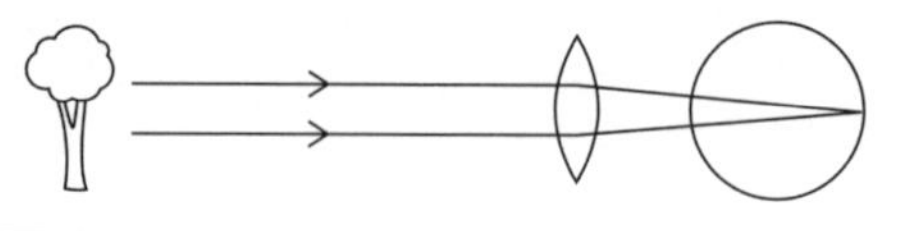
C.	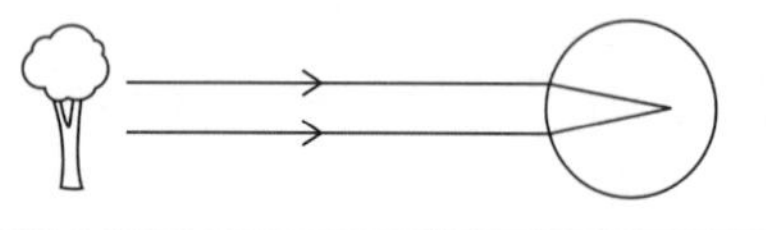	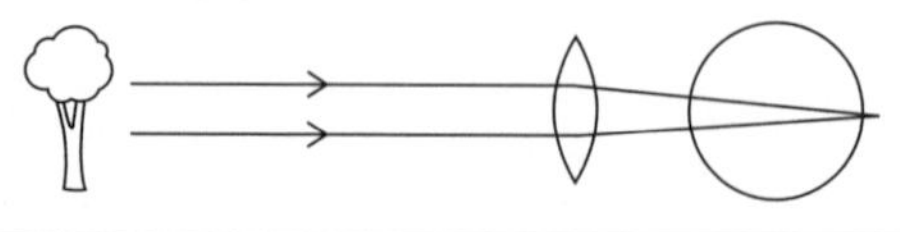
D.	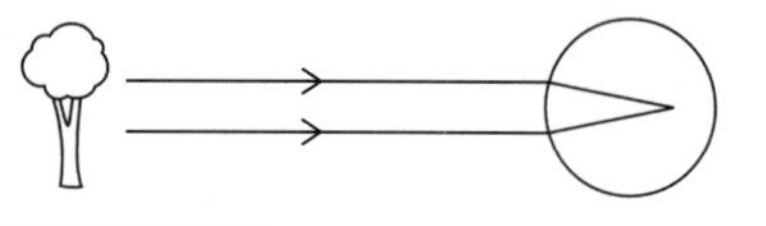	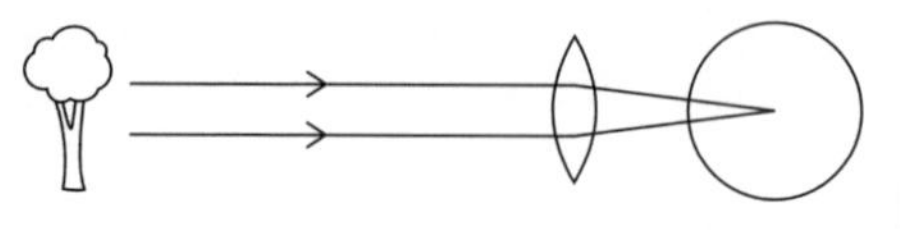

20 Renal dialysis involves passing blood from a patient past a dialysate solution in order to remove waste such as urea from the blood.

Which diagram correctly shows possible concentrations of urea and the direction of flow of both solutions in a dialysis machine?

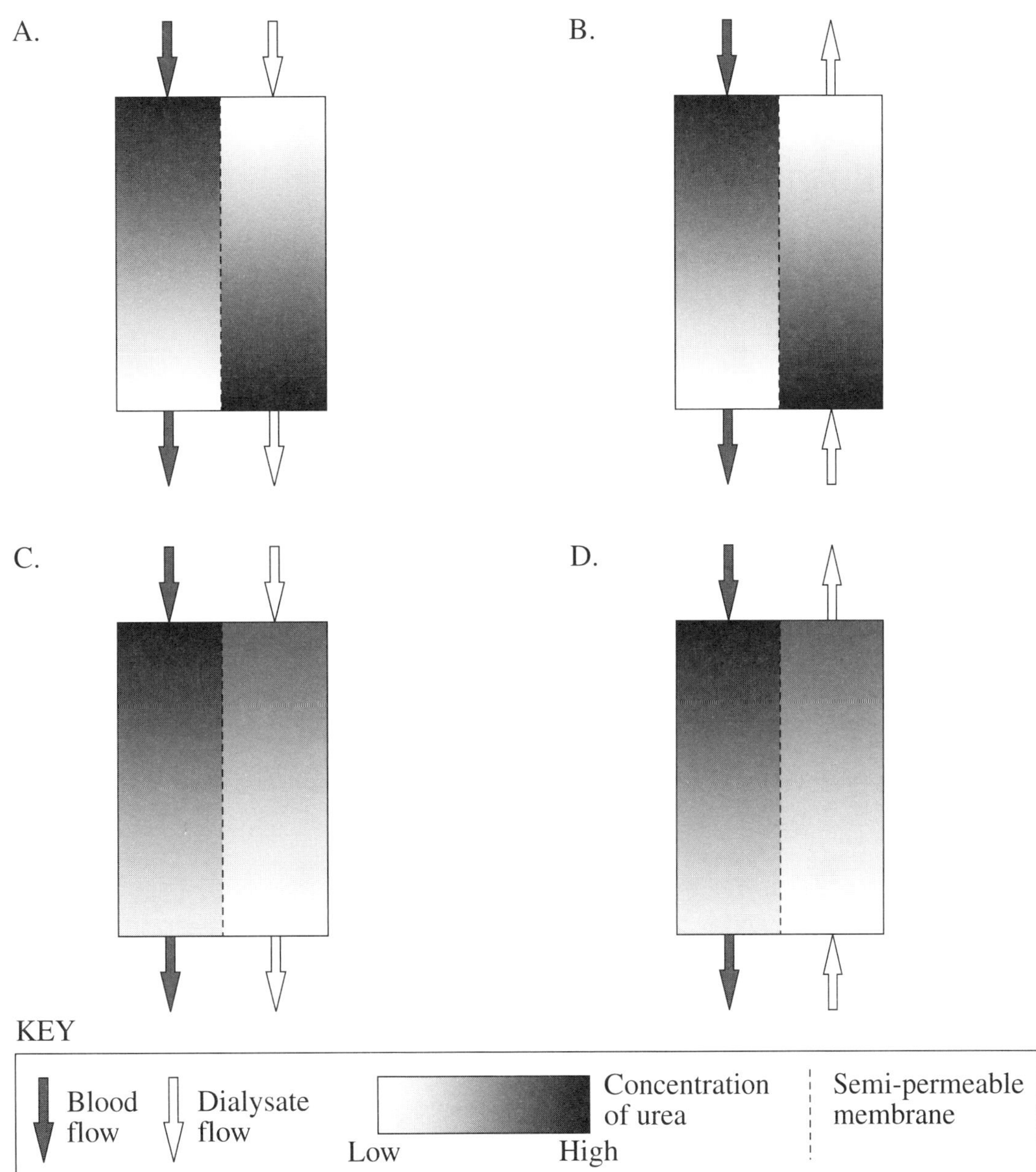

2022 HIGHER SCHOOL CERTIFICATE EXAMINATION

Centre Number

Student Number

Biology

Section II Answer Booklet

80 marks
Attempt Questions 21–32
Allow about 2 hours and 25 minutes for this section

Instructions

- Write your Centre Number and Student Number at the top of this page.
- Answer the questions in the spaces provided. These spaces provide guidance for the expected length of response.
- Show all relevant working in questions involving calculations.

Please turn over

Question 21 (5 marks)

(a) Outline ONE way that a pathogen can pass from person to person. **2**

....

....

....

....

(b) The following key can be used to classify some pathogens. **3**

Complete each empty box with an appropriate pathogen.

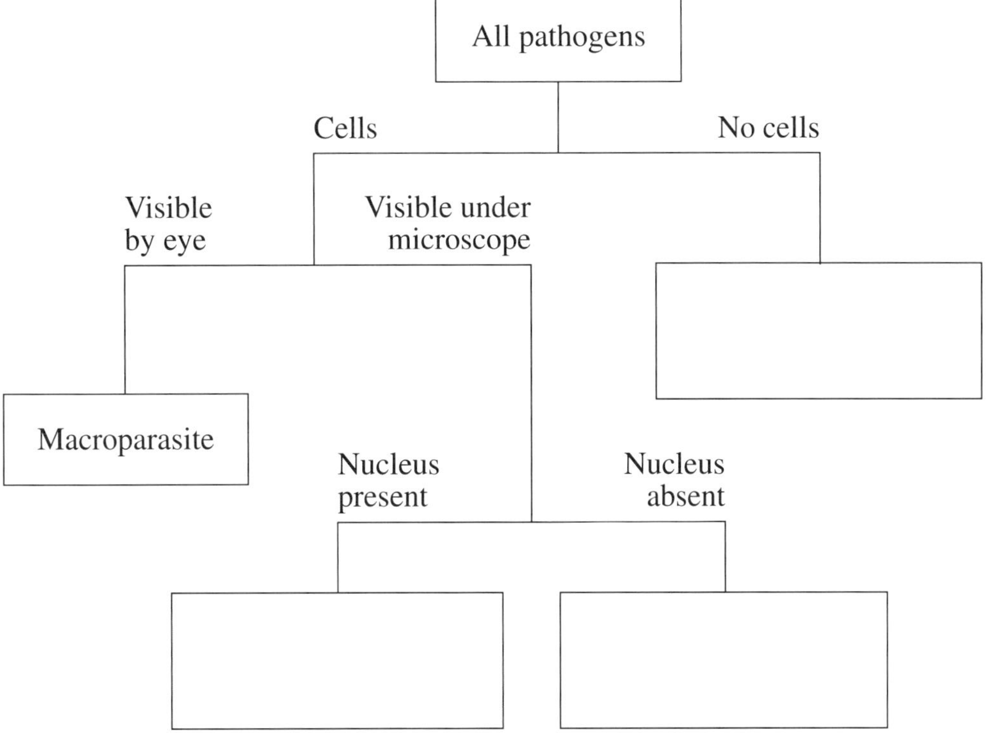

Question 22 (3 marks)

Eggplant fruit comes in three colours: dark purple, white and violet. A genetic cross between the dark purple and white eggplants will always result in the violet phenotype. **3**

What phenotypic ratio would you expect to see when two violet offspring are crossed? Show your working.

Question 23 (4 marks)

(a) Outline the process of artificial pollination. **2**

....

....

....

....

(b) Explain a possible outcome of the use of artificial pollination on subsequent populations. **2**

....

....

....

....

....

Please turn over

Question 24 (8 marks)

Birth defects in humans can be caused by chromosomal abnormalities.

Maternal age (years)	*Prevalence of chromosomal abnormalities* (per 1000 births)
20	1.5
30	3
35	8
40	22
45	38

(a) Draw a suitable graph of the data provided in the table. Include a suitable line of best fit. **3**

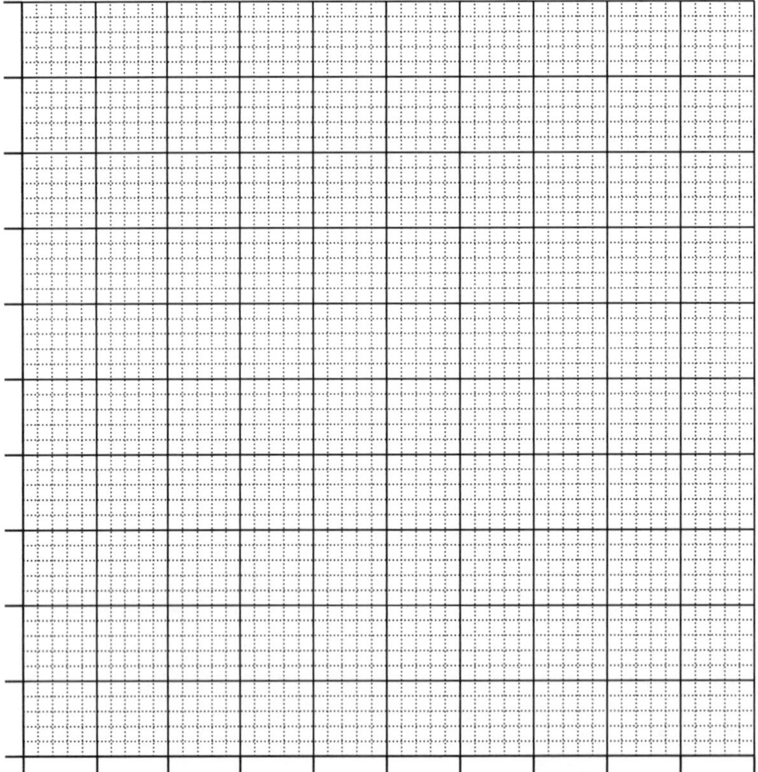

Question 24 continues on the following page

Question 24 (continued)

(b) Outline the trend shown in the data. **2**

....

....

....

....

(c) Explain the cause of a type of chromosomal mutation. **3**

....

....

....

....

....

....

....

....

....

....

....

....

End of Question 24

Question 25 (6 marks)

Offspring are rarely identical to their parents. **6**

Compare the ways in which genetic variation can arise from asexual and sexual reproduction.

Question 26 (5 marks)

Jelly Bush honey has been used by Aboriginal Peoples to treat cuts, sores and burns. Recent studies have shown that Jelly Bush honey has a very high level of methylglyoxal which is known to help fight infection. **5**

A scientist wants to test the effectiveness of Jelly Bush honey using agar plates.

Design a safe procedure that the scientist could use in a laboratory to investigate the effectiveness of Jelly Bush honey as a pharmaceutical to inhibit bacterial growth.

Question 27 (6 marks)

The incidence of cervical cancer in the Australian population is shown in the graph from 1985–2015. Public health campaigns commenced during this period to reduce the incidence of cervical cancer. This included a national screening program to detect pre-cancerous cells and a vaccination program against human papillomavirus (HPV) which causes most cervical cancers. **6**

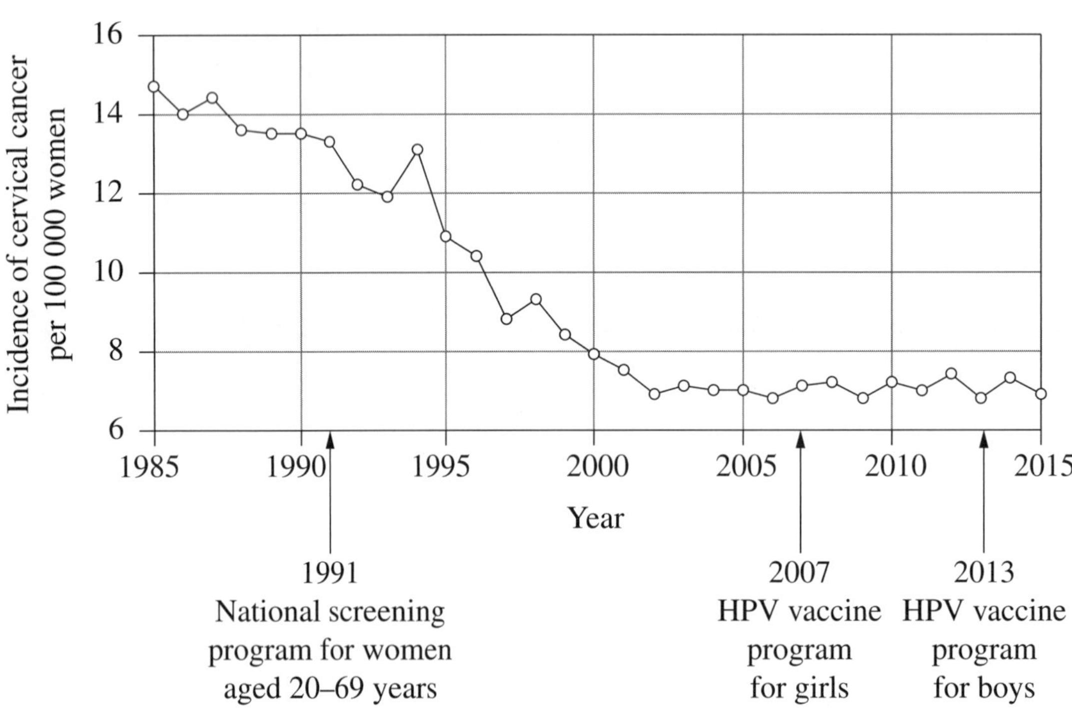

Evaluate the success of these campaigns in reducing the incidence of cervical cancer in Australian women. Include reference to the data in your answer.

....

....

....

....

....

....

....

....

....

....

Question 27 continues on the following page

Question 27 (continued)

....

....

....

....

....

....

....

....

....

....

....

End of Question 27

Please turn over

Question 28 (7 marks)

Soon after the discovery of the structure of DNA, scientists began investigating how DNA replicated. Three models were proposed.

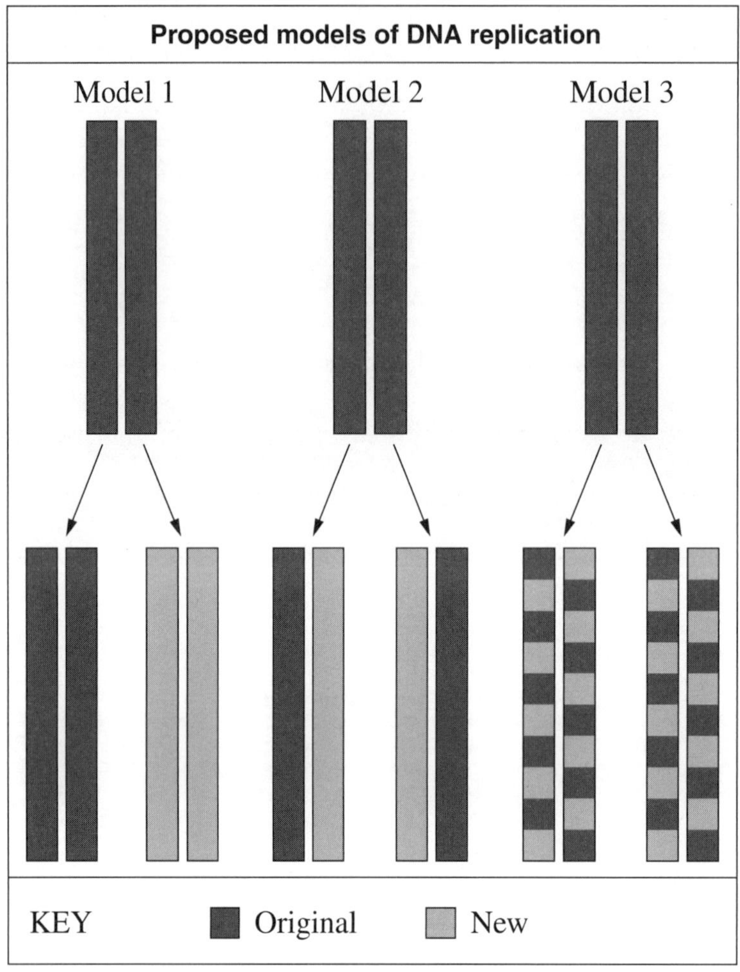

(a) Identify which of the models above is currently accepted. **1**

....

Question 28 continues on the following page

Question 28 (continued)

(b) Describe the process of DNA replication. **3**

....

....

....

....

....

....

....

....

....

....

....

(c) Outline the ways in which the DNA of prokaryotes and eukaryotes differ. **3**

....

....

....

....

....

....

End of Question 28

Question 29 (7 marks)

Bt cotton has been genetically engineered to produce an insecticide that kills cotton bollworm. It was introduced to a cotton-producing nation in 2002.

The graphs show trends of national cotton yield, % Bt cotton grown, total insecticide use, insecticide use to control bollworms and insecticide use to control another insect pest (hemiptera).

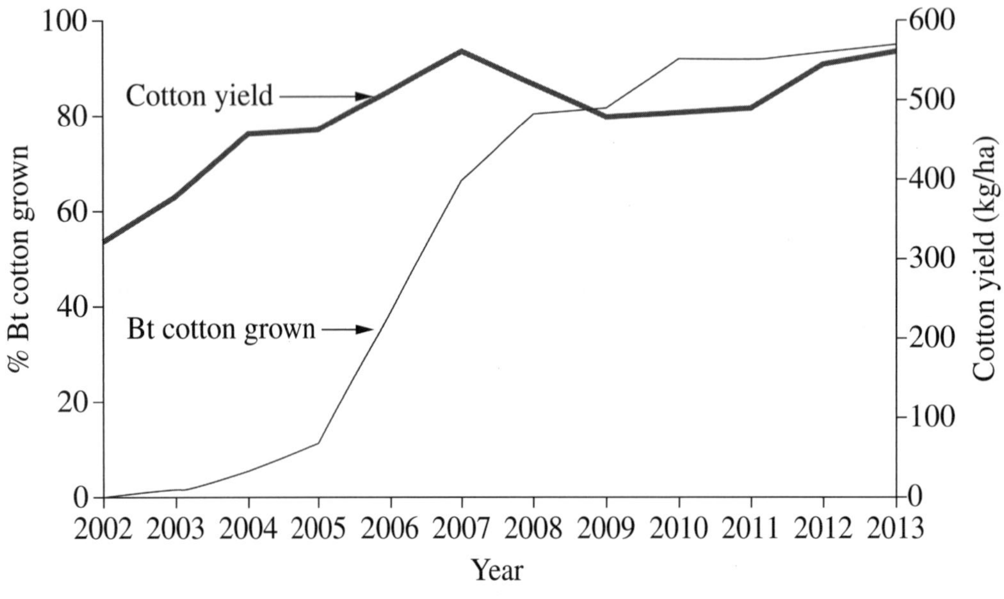

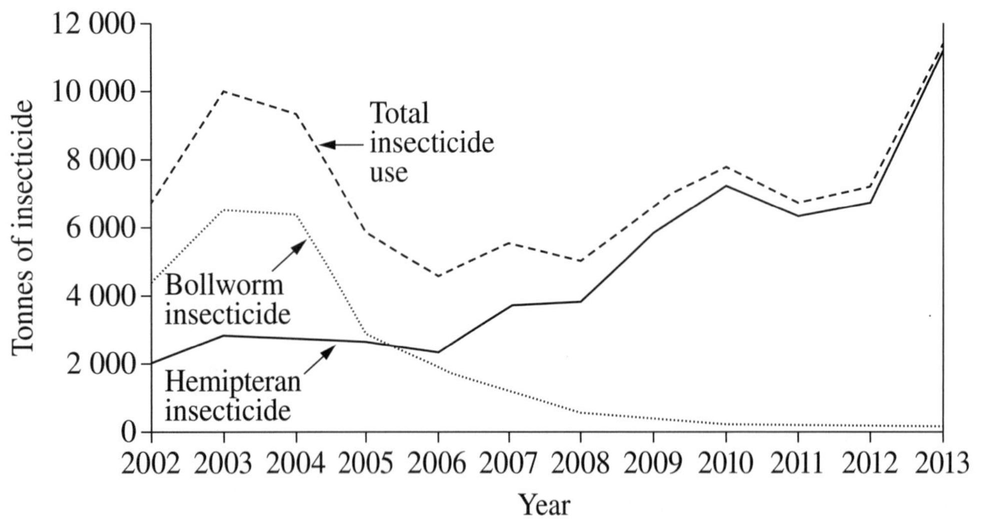

(a) Explain ONE reason why cotton yield changed between 2002 and 2013. **2**

....

....

....

....

Question 29 continues on the following page

Question 29 (continued)

(b) To what extent do the data support the use of Bt cotton as a method of disease control in cotton? **5**

End of Question 29

Question 30 (7 marks)

Malaria is a disease transmitted by a mosquito vector. There has been no effective vaccine developed. 7

Refer to the maps below to answer Question 30.

Map 1 **Global distribution of the malaria-transmitting mosquito**

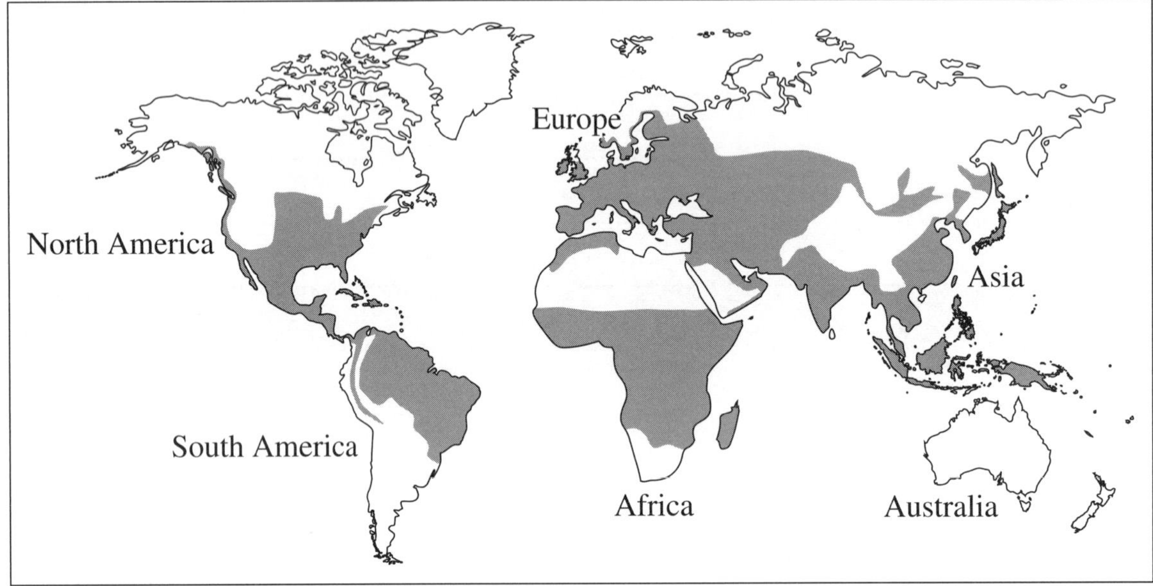

Map 2 **Global distribution of malaria cases**

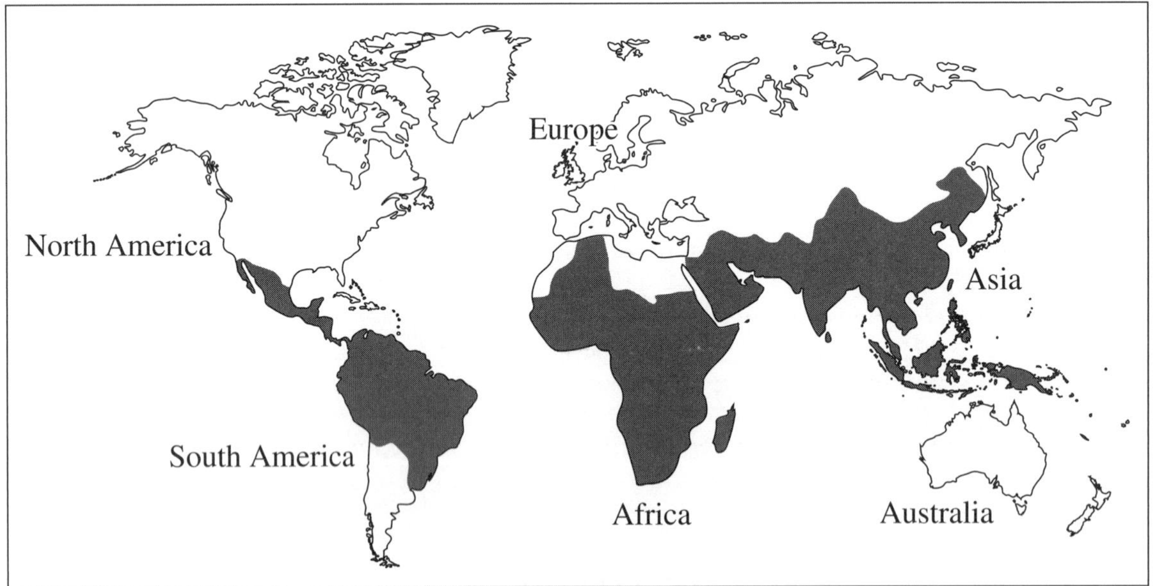

Question 30 continues on the following page

Question 30 (continued)

Discuss possible reasons for the differences in the distribution of malaria and its vector. Include detailed reference to Map 1 and Map 2.

End of Question 30

Question 31 (15 marks)

Studies have shown that lung cancer can be linked to environmental causes.

In one historical study across 29 health districts in Japan, non-smoking married women aged 40 and above were followed up for 14 years (1966–79) and annual mortality rates for lung cancer were assessed according to the smoking habits of their husbands. They were compared to women who smoked.

The results are shown.

Group	*Number of women*	*Lung cancer mortality rate* (per 100 000)
Non-smoker women with non-smoker husbands	21 895	8.7
Non-smoker women with smoker husbands	69 645	15.5
Women who smoke	17 366	32.8

(a) (i) Evaluate the method used in this epidemiological study. **4**

....

....

....

....

....

....

....

....

....

....

....

....

Question 31 continues on the following page

Question 31 (continued)

(ii) Justify conclusions that could be drawn from the results of the study. **3**

....

....

....

....

....

....

....

....

....

Question 31 continues on the following page

Question 31 (continued)

(b) Lung cancer can be linked to genetic causes. One of the genes frequently studied in lung cancer tissue is the Epidermal Growth Factor Receptor (EGFR) gene. It codes for EGFR protein, which is composed of one polypeptide chain.

(i) Construct a flow chart to outline the synthesis of the EGFR protein from the EGFR gene. **4**

Question 31 continues on the following page

Question 31 (continued)

The structure of the EGFR protein includes a receptor and an enzyme component. The function of the protein is to help the cell to regulate cell division.

EGFR mutations are present in about 32% of cases of Non-Small Cell Lung Cancer (the most common type of lung cancer).

(ii) Explain how a mutation in the EGFR gene could result in changes in protein structure and function to increase the risk of lung cancer. **4**

....

....

....

....

....

....

....

....

....

....

....

....

....

....

....

End of Question 31

Question 32 (7 marks)

Researchers have identified a gene that determines the inflammatory response of lung cells to infection with a virus. An allele of this gene is associated with increased inflammation and increased chance of death from the virus. 7

The table shows the percentage presence of the allele in people with different ancestries.

Ancestry	*Percentage of population with the allele*
South Asian	60.3
European	15.1
African	2.4
East Asian	1.8

Explain how mutation, natural selection, genetic drift and gene flow could have led to these differences in the gene pools of populations with differing ancestry.

....

....

....

....

....

....

....

....

....

....

....

....

....

....

....

Question 32 continues on the following page

Question 32 (continued)

End of paper

2022 HSC Examination Paper

Sample Answers

Section I (Total 20 marks)

1 B Vasodilation allows the blood to circulate close to the skin surface, facilitating loss of heat.

2 C Breaking down body fat is functional and therefore physiological.

3 A Antibodies are protective proteins produced by the immune system in response to the detection of antigens.

4 B Redness, swelling, increased blood flow and heat are inflammatory responses to injury.

5 B Each category of infectious disease is discrete so best represented as discrete columns.

6 D Z represents the uterus where implantation normally occurs.

7 D The nucleus from the somatic cell must be inserted into an egg that no longer has its nucleus (enucleated egg).

8 B Muscle cells are somatic cells so they would contain 20 pairs of chromosomes (40 chromosomes in total).

9 D The line graph for those over 65 years of age is consistently below that of the younger age groups.

10 B The independent variable is the one that is changed during the experiment and is not influenced by any other variable in the experiment. In this case Pasteur broke the flask neck hence flask is the independent variable.

11 A The best explanation for the unsealed broth becoming spoilt is that cells were able to contaminate the exposed broth and then reproduce.

12 A X is deoxyribose sugar and Y is the phosphate group that together form the DNA backbone. Z must be thymine as it forms two hydrogen bonds with its pairing purine base, adenine.

13 C After several replication cycles both nucleotides of the base pair have changed.

14 A As the parents of the affected daughter were unaffected, the condition is recessive. If the condition was carried as an allele on the X chromosome and hence sex-linked, the male parent of the affected daughter in the third generation would also have had to be affected.

15 C Red is dominant so one parent was heterozygous red while the other was homozygous red.

16 C Non-coding DNA is not directly used for polypeptide synthesis but may impact gene regulation and expression.

17 B During meiosis, homologous pairs line up along the metaphase plate to enable separation in anaphase. The homologous chromosome pairs should be the same size and structure.

18 A Somatic mutations are not passed on to the offspring. As two of the three offspring of individual 2 are affected it is likely the mutation occurred in that individual's germ-line cells.

19 D Myopia results in an image forming in front of the retina. A convex lens will further shorten the distance for image formation.

20 D The dialysate solution relies on osmosis so it initially contains no urea. Dialysis will be most effective when the blood and the dialysate have the greatest difference in urea concentration throughout their exposure.

Section II

Question 21 (Total 5 marks)

(a) One way that a pathogen can pass from person to person is through mucus build up in the nasal passages of the host. Coughing and sneezing can transfer mucus containing the pathogen to the air and then to the next host through inhalation. (*2 marks*)

(b)

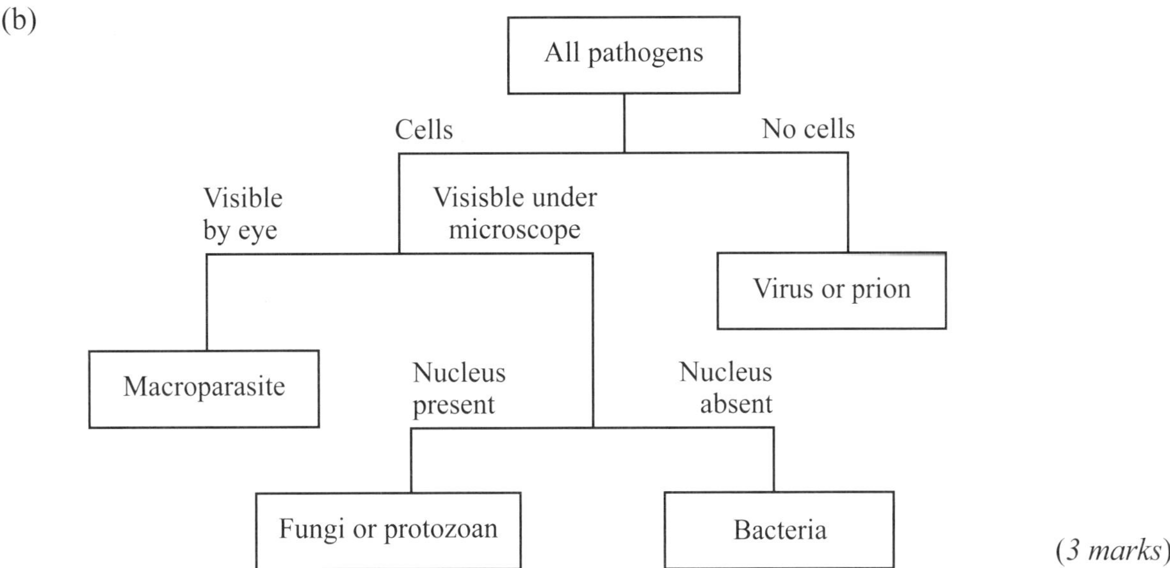

(*3 marks*)

Question 22 (Total 3 marks)

Violet × Violet

Parent 1

Parents' gametes	→↓	V	v
Parent 2	V	VV	Vv
	v	Vv	vv

1 purple : 2 violet : 1 white

This incomplete dominance cross will result in offspring with a phenotypic ratio of 1 purple: 2 violet: 1 white. (*3 marks*)

Question 23 (Total 4 marks)

(a) In artificial pollination pollen is collected from the anthers on the stamens of flowers and transferred by humans to the stigmas of specific flowers to allow pollination and fertilisation. Steps are taken to ensure accidental pollination from unintended sources does not occur. In some cases, such as self-pollinating rice and maize, the stamens or tassels have to be removed (emasculated) to ensure natural pollination does not occur. *(2 marks)*

(b) A possible outcome of artificial pollination is improved plant production of fruit or seeds especially if there is a shortage of naturally occurring insect pollinators. Naturally occurring pollinators may be lacking because some crops are grown in glasshouses (to extend the growing season or promote optimal growing conditions) or in areas where the number of insect pollinators has declined because of widespread use of insecticides. *(2 marks)*

Question 24 (Total 8 marks)

(a)

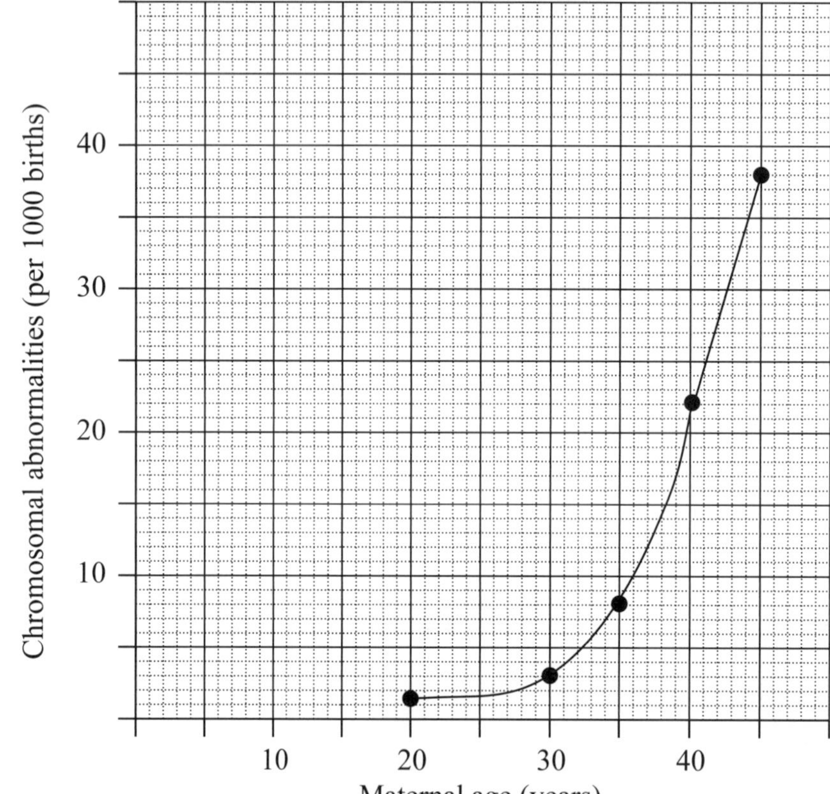

(3 marks)

(b) The trend shows there is an increase at an exponential rate in the prevalence of chromosomal abnormalities with increasing maternal age. The rate of increase in chromosomal abnormalities is much greater between the ages of 35 and 40 than between 30 and 35. *(2 marks)*

(c) Chromosomal mutations occur during cell division. An entire chromosome can be lost or gained as a result of non-disjunction. Non-disjunction occurs when the homologous chromosomes fail to separate or segregate at Anaphase. As a result, one of the daughter cells has an additional chromosome and the other has a missing chromosome. When non-disjunction occurs during meiosis this impacts the gametes and can result in aneuploidy. Non-disjunction rates increase with increasing maternal age. *(3 marks)*

Question 25 (Total 6 marks)

Asexual reproduction results from mitotic cell divisions in which the daughter cells are genetically the same as the parent cell. Offspring are genetically the same as the single parent. Asexual reproduction is an important form of reproduction in many unicellular, colonial and simple multicellular organisms, bacteria, protists, fungi and plants. It facilitates rapid population growth, expansion and dispersal but results in minimal genetic variation. The only source of genetic variation in asexual reproduction is somatic cell mutations. Some phenotypical variation may result from varying environmental factors.

In contrast there are numerous causes of genetic variation in sexual reproduction. These include the following:

- A zygote results from fertilisation. This combines half the genetic material in the form of a gamete from one parent with half the genetic material from another parent. Through mitotic cell divisions the zygote develops into an offspring containing unique combinations of genes.
- Meiosis is the essential cell division process that results in the production of gametes. Within meiosis the processes of crossing over and random assortment of homologous chromosomes promote genetic diversity. Crossing over between adjacent or non-sister chromatids results in the exchange of genetic material and new, unique combinations of genes on particular chromatids. Random assortment of the members of the homologous pairs adds greatly to genetic diversity in gametes that contain unique combinations of chromosomes.
- Germ-line mutations are an additional potential source of genetic variation in sexual reproduction.

The wide diversity of angiosperms, gymnosperms, vertebrate and invertebrate animals results from their reproduction sexually. (*6 marks*)

Question 26 (Total 5 marks)

Safe procedures include:

- inoculation of the agar plates using a sterile inoculating loop or needle and a culture of bacteria that are not pathogenic to humans
- no opening of the agar plate after inoculation, inversion and labelling; and after incubation and assessment of the number of bacterial colonies, disposal of plates by autoclaving.
- disinfecting hands, work benches and areas before and after use
- using personal protective wear such as gloves, laboratory coat and eye protection
- no eating or drinking in the laboratory.

Reliable procedures include:

- ensuring a large number of agar plates are used so results are statistically reliable.

Procedures to ensure a valid investigation include:

- ensuring that a sufficient number of agar plates are controls with no Jelly Bush honey and which are used to compare the number of bacterial colonies with the plates that contain Jelly Bush honey
- ensuring all variables except the use of Jelly Bush honey are controlled with the same exposure to bacteria, the same incubation and observation and the same recording of results (number of bacterial colonies)
- repetition of the investigation by varying the concentrations of Jelly Bush honey used on the experimental agar plates to ensure the effectiveness of the honey is really tested.

Aim: to test the effectiveness of Jelly Bush honey as an antibacterial

Method:

1. Prepare 10 agar plates containing Jelly Bush honey and 10 agar plates without.
2. Inoculate the 20 plates using an inoculating loop and non-pathogenic bacteria.
3. Invert the plates, label and incubate at 30 degrees for a week, observing at regular intervals. Do not open the plates. Record the number and size of bacterial colonies that develop.
4. Autoclave plates.
5. Compare the results to draw a conclusion. *(5 marks)*

Question 27 (Total 6 marks)

In evaluating the success of the national screening program for women aged 20 to 69 years it must be noted that prior to the program the trend in incidence of cervical cancer in women was gradually declining. Between 1991 and 2002 the rate of decline of the incidence dramatically increased. A brief spike in the incidence in 1994 could probably be explained by the increase in detection as a result of more women coming on board with screening rather than an actual increase in the incidence of cervical cancer. It can be assumed that the detection of pre-cancerous cells led to treatments that prevented the onset of cervical cancer in about 4 women per 100 000 over the 11-year period.

Therefore the data suggests the screening program was a great success (dropping from over 13 per 100 000 to 7 per 100 000) but that its benefits plateaued in about 2002 when the incidence reached 7 women per 100 000. This conclusion could be made with greater confidence if the number of women screened was reported and the stages of the cancer that were detected were also revealed.

Evaluating the success of the vaccination programs appears to be premature. In the eight years in which data was recorded after the vaccination of girls commenced, the incidence has not greatly changed. This would be expected, as very few of the girls would have been regularly screened by 2015. The benefits of vaccination for boys would be expected to take many more years before transmission levels of the HPV decline sufficiently to reduce the incidence of cervical cancer in women. Collection of data validating the reduction of the occurrence of HPV as a result of the vaccination programs would be of benefit in being able to evaluate the success of the vaccination sooner than can be expected from detecting incidence of the cancer. *(6 marks)*

Question 28 (Total 7 marks)

(a) Model 2 (*1 mark*)

(b) DNA replication:

1. The DNA double helix uncoils.
2. The double strands separate as the hydrogen bonds between the matching base pairs are broken by the enzyme helicase.
3. Primers (short segments of RNA) bind with the open strands, depending on whether the stand is the leading or lagging strand.
4. Nucleotides slot into the leading strand continuously from the 3′ end to the 5′ end.
5. Nucleotides slot into the lagging strand in short segments from the 5′ end to the 3′ end interspersed with the primers. DNA polymerase enzymes promote these two elongation processes (steps 4 and 5).
6. An enzyme removes RNA primers and the vacant slots are filled with nucleotides.
7. Enzymes check for errors and correct. They ensure that adenine is paired with thymine and cytosine with guanine.
8. A special enzyme catalyses the synthesis of the repeating segments of DNA tips (telomeres).
9. Each double strand re-coils into a double helix that consists of one template from the original molecule and one new strand. (*3 marks*)

(c) Prokaryotes do not have a nucleus so their DNA largely consists of a single circular supercoiled chromosome located in a region called the nucleoid. Some DNA may be found as small circular plasmids that can self-replicate. There is only one copy of each gene. Replication is fast proceeding around the circle in opposite directions. Small amounts of protein are associated with the DNA to help form the supercoiling. Most DNA is coding.

Eukaryotes have DNA in the nucleus found as chromatin. Chromatin is a macromolecule complex consisting of the protein histone and DNA. During mitosis and meiosis the DNA packs densely as linear chromosomes. There are two copies of each gene. DNA is also located in mitochondria and chloroplasts where it is often circular and capable of replication. DNA replication is relatively slow. Most DNA is non-coding. (*3 marks*)

Question 29 (Total 7 marks)

(a) The cotton yield increased between 2002 and 2013 because there was an increase in the percentage of genetically modified cotton grown. This GMO cotton was more productive because it resisted attack by bollworms. (*2 marks*)

(b) The data shows that the yield of cotton in kg/ha increased over the 11-year period and the use of bollworm insecticide declined. These figures would support the use of Bt cotton from the economic and disease-caused-by-bollworm perspectives.

As a method of disease control in cotton, the data suggests that while one disease was controlled the hemipteran insect pest dramatically increased. As far as disease control is concerned, this led to an increase in the total insecticide use as a result of greatly increased use of hemipteran insecticide. Since 2007 the cotton yield initially declined, then levelled and finally increased slightly. The data suggests that strategies for disease control need to consider the full range of pests and diseases to be targeted to ensure that eliminating one does not result in other pests and diseases proliferating.

In conclusion, the data supports the use of Bt cotton as a method of disease control in the short term but not in the long term. (*5 marks*)

Question 30 (Total 7 marks)

The comparison of the maps shows that the malaria vector is widespread in areas of England, Europe, the Middle East, Japan and North America but the prevalence of the disease in these areas is not high. Conversely, there are regions in central Asia, the Arabian peninsula, and southern and north-western Africa that are not exposed to the vector but have a high incidence of malaria.

It can be assumed that the regions exposed to the malaria-transmitting vector but not experiencing the disease have developed methods of prevention of transmission and disease control. The elimination of the pathogen of the disease results in a vector no longer transmitting the pathogen and hence the disease remains absent. Prevention could result from:

- use of effective anti-malarial drugs such as chloroquine
- populations that have high natural immunity to the pathogen
- mosquito-proof housing, clothing and repellants
- pesticide use, irrigation and drainage management of the environment in high population-density regions and breeding locations
- education programs to avoid activities at times and locations when the vector is active.

Disease control could result from the introduction of genetically modified vectors that lack the ability to transmit the pathogen.

Variations in the gene pool of populations may also help explain why some areas do not experience malaria despite exposure to its vector. People who are heterozygous for sickle-cell anaemia are reported to be resistant to the pathogen.

Regions that have a high incidence of malaria yet are not exposed to the vector could be the result of a highly mobile human population and urbanisation. Not only may sufferers be mobile but also goods such as car tyres may carry water contaminated by vectors into otherwise unexposed regions. Some regions with a high incidence of malaria may be explained by variations in the population demographics. A high proportion of young people may reduce the effectiveness of educational programs or experience low levels of natural immunity. (*7 marks*)

Question 31 (Total 15 marks)

(a) (i) There are many limitations of this epidemiological study on the ability to draw conclusions from the results. Epidemiological studies correlate variables but do not prove cause and effect.

One concern about the method used in this study is the use of mortality rates as the basis of comparison of the incidence of lung cancer in three groups of women. Comparing the morbidity rates of lung cancer would be a better indicator of the incidence of the disease rather than the number of deaths from the disease.

The large numbers of women studied in each group increases the reliability of this study.

The study was carried out over 29 health districts but there is no information about the size or distribution of the districts. There may be other variables that were significant relating to lung cancer that occurred in the districts.

There is insufficient information about any attempt to control variables such as exposure to mutagens, including air pollution, access to medical care and age (apart from that they are over 40 and married) to indicate whether the study was valid.

Within the two groups containing smokers there could be a great variation in the number of cigarettes smoked, the type of household ventilation and hence the actual levels of smoke inhaled by the women. The third group of smoking women may have had either smoking or non-smoking husbands so there could be significant variation in their exposure levels to cigarette smoke.

This study would be improved if the duration of the investigation was greatly increased. This is made worse because mortality rates, not morbidity rates, were compared and lung cancer may take many years to result in death. (*4 marks*)

(ii) It could be concluded that rates of death from lung cancer in the populations studied increase greatly as a result of smoking and moderately by being exposed to secondhand cigarette smoke.

This can be justified because there is almost four times the mortality rate for smoking women versus non-smoking women who are not married to a smoker (and presumably not exposed to secondary smoke) and twice that when smoking women are compared with non-smoking women with a smoking husband. The large number of participants in the study over numerous locations adds to justification of the conclusions.

Justification for conclusions that could be drawn from this study could result from reliability with numerous other studies that have successfully correlated lung cancer with smoking. (*3 marks*)

(b) (i)

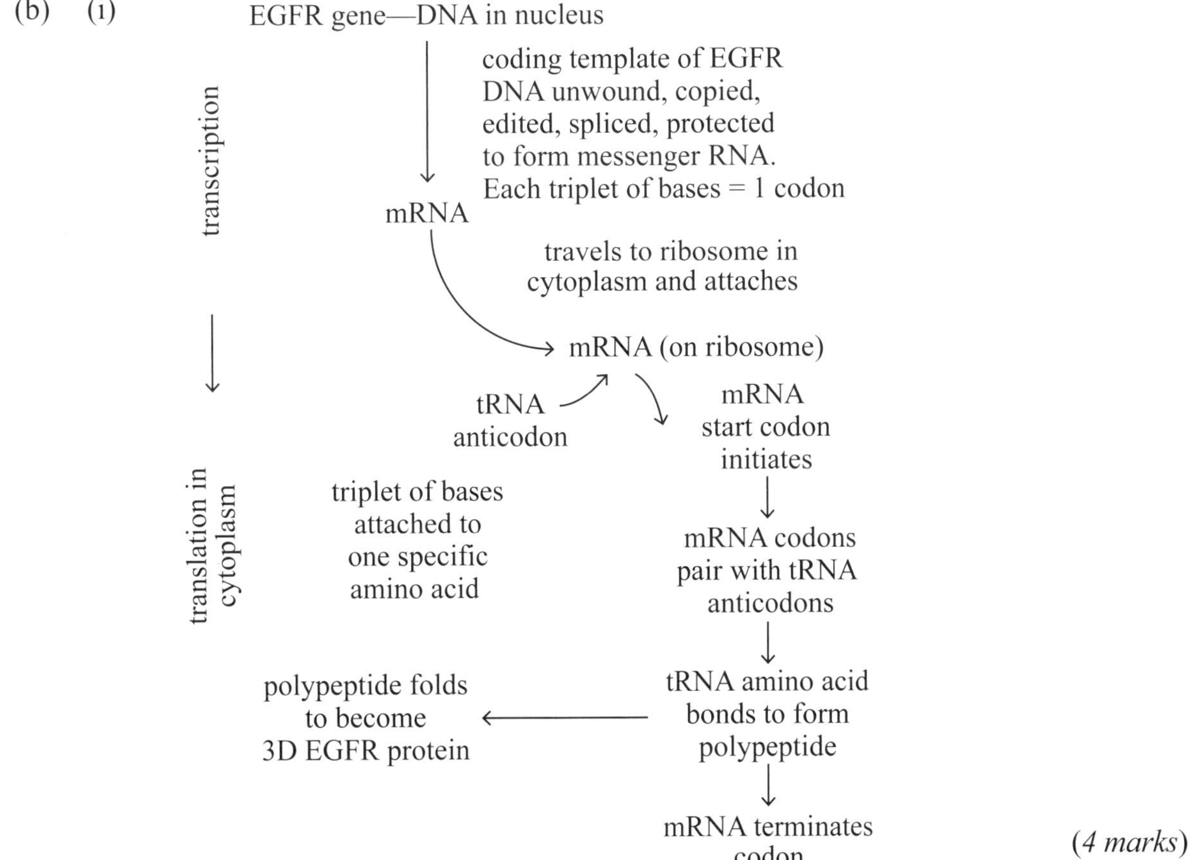

(*4 marks*)

(ii) Lung cancer is a result of uncontrolled cell division of the lung tissue. Cell division is normally regulated.

A base substitution mutation could result in a polypeptide chain with a substituted amino acid. This change in the primary structure might impact the secondary structure of the protein, such as the formation of beta pleated sheets or alpha helices. Irrespective of whether there are any regular secondary shapes, a polypeptide such as an EGFR contort to form a regular and predictable three-dimensional shape. While hydrogen bonds may weakly hold some aspects of the macromolecule in place, other covalent bonds such as disulfide bridges maintain a stable structure. Other regions along the polypeptide chain may have properties that make them attract or repel water. The resulting tertiary structure of a polypeptide such as EGFR is integral to its normal functioning. A change in just one amino acid in a polypeptide has the potential to change the secondary and tertiary structure. Other types of mutation have the potential to alter the structure of the polypeptide even further or prevent its formation altogether.

EGFR contains an enzyme component. This means the protein acts to promote and control a chemical reaction. Enzymes often act by having an active site that fits with a substrate to alter the activation energy required for the reaction to occur. The physical shape of the active site is vital to its functioning. A mutation could render the enzyme component of the protein non-functional by changing its shape.

As EGFR has both a receptor and an enzyme component it could be assumed that part of the protein receives a signal (electrical or chemical) and responds, possibly through regulation of the activity of the enzyme component. If the structure of the receptor component is altered, it may lose the ability to regulate. (*4 marks*)

Question 32 (Total 7 marks)

Mutation is the ultimate source of genetic variation in the gene pool of a population. It is unlikely that mutation alone could explain the different percentages (60.3% in South Asia versus 1.8% in East Asia).

If a particular allele increases the chances of an individual surviving and reproducing, natural selection will occur to increase the allele frequency in that population. In the case of the allele associated with increased lung inflammation resulting from viral infection, its impact is deleterious. It could be assumed that in areas where there is a high rate of viral lung infection, natural selection would operate to reduce the prevalence of this allele. Ancestors of people from East Asia and Africa would have had greater exposure to the lung virus than Europeans; South Asians would have experienced even less. Alternatively, the allele may have remained in high frequency in areas where medical treatments have outweighed the impact of natural selection. South Asia may have excellent medical facilities so that the impact of the allele increasing the risk of death from viral lung infections is negated by the level of care. Carriers of the allele may no longer have reduced life expectancies and reproduction rates so the allele remains in high frequency.

It is possible that the allele identified as being associated with increased inflammation and greater risk of death from lung viral infections may have an unidentified benefit. This is the case with the condition sickle-cell anaemia. The homozygous condition for the allele is generally fatal but the heterozygous condition promotes resistance to malaria so that the allele has remained frequent in populations exposed to the malarial parasite.

Random chance events or genetic drift occur in small populations and may also positively or negatively affect the gene pool. Such random events include processes that are involved in gamete production and fertilisation of gametes. Genetic drift generally has a minor impact on the size and diversity of the gene pool. For genetic drift to explain some of the differences in the frequency of the allele associated with increased inflammation, its impact must have occurred very early in the evolution of the ancestral populations of the identified areas when the populations were still very small.

Gene flow is the result of migration between populations. Generally immigration results in an increase in diversity of the gene pool. Gene flow may be considered to be more likely to occur between neighbouring populations such as East and South Asia. The data from the allele frequency table does not support the idea that significant gene flow has occurred between these two regions. Gene flow being linked with close proximity of populations may not always be the case, as in recent times large movements of refugee populations have resulted from political unrest and violence.

The table identifies large differences in allele frequency across the four identified populations. The most likely explanation of some of these differences is mutation followed by natural selection and gene flow. (*7 marks*)

Acknowledgements

Summarised text, Q26, 2006 HSC (Q19, Module 7), S Pincock, 2005, Nobel Prize winners Robin Warren and Barry Marshall, *The Lancet*, 366(9495): 1429

Diagram, Q9, Module 6, National Human Genome Project, 2006

Diagram, Q10, Module 6, J Groth, 2006, Reproductive and therapeutic cloning

Graphs, Q11, 2009 HSC (Q7, Module 8), M Jones and G Jones, 1997, *Advanced Biology*, Cambridge University Press

Extracted text, Q33(e), 2013 HSC (Q25, Module 5), 'DNA kits can bring unwanted surprises', Philip Sherwell, *The Telegraph*, 7 January 2013

Notes

Notes